Praise for

IF FOOD COULD TALK

"A charming wake up call and an introduction to thinking beyond your plate. Learn about the origin of some favorite foods, how to deal with the threat climate change poses to them, AND celebrate with some special recipes. Truly food for thought."

—Dr. Thomas Lovejoy

President of the Amazon Biodiversity Center, a Senior Fellow at the United Nations Foundation and university professor in the Environmental Science and Policy department at George Mason University, "godfather of biodiversity"

"Whoa—there's 7,500 types of apples, archeologists have found fossilized cherry pits in caves, there was actually a chocolatier named Nestle. Who knew? And who knew that apples, cherries, chocolate and more are endangered by global warming? In history's only book, I suspect, that is half jeremiad and half cookbook, *If Food Could Talk* teaches about foods at risk. The book is fascinating and accessible. Prof. Dumas is both an accomplished scientist and teacher, exploring the law of unexpected consequences of unexpected importance."

—Robert Sapolsky

Author of *Behave: The Biology of Humans at Our Best and Worst*

"Dr. Dumas's book brings climate change home in a way that will resonate with readers. By describing foods that everyone is familiar with and a few that are beloved (e.g. chocolate), he brings home the history and cultural significance of these foods, while illustrating the perils posed by climate change. Even more, by adding recipes, he turns the esoteric into the prosaic, making us not just understand their history, but how they taste at the dinner table.

"No disrespect to polar bears, but here is a book that personalizes what climate change means using a medium that we all understand—food. Not just any food, but the best of what nature has to offer, from wine, to chocolate to avocados. Dr. Dumas provides insight into their background and culture and, more to the point, how each food will be impacted by climate change. This is a unique and personal view of climate change that everyone who eats will appreciate."

—Lewis Ziska, PhD

Assoc. Professor, Mailman School of Public Health, Columbia University

"An important and delectable call to action on our most precious human resource. Sounds the alarm and gives a powerful pathway for our species to eat responsibly today and into the future."

—Josh Tickell

Author of *Kiss the Ground*

If Food Could Talk:

Stories from 13 Precious Foods Endangered by Climate Change

By Theodore Constantine Dumas

ISBN 978-1-64663-239-8

Published by

köehlerbooks™

210 60th Street
Virginia Beach, VA 23451
800–435–4811
www.koehlerbooks.com

IF
FOOD
COULD
TALK

STORIES FROM 13
PRECIOUS FOODS
ENDANGERED BY
CLIMATE CHANGE

THEODORE
CONSTANTINE
DUMAS

VIRGINIA BEACH
CAPE CHARLES

Images were captured by Nhu N. Dumas and Theodore C. Dumas

Greenhouse illustration was created by Theodore C. Dumas

CO_2 emissions pie chart was created by Dr. Sultan Al-Salem
(cited in the text and the References list)

I DEDICATE THIS BOOK to my family, immediate and extended. Thank you!

To my yiayias and pappous for immigrating to the United States from Greece. Nothing against Greece. It's a great place to visit, but it's not the United States, the "land of golden opportunities," as my grandparents would often say. None of my grandparents were highly educated, but they had the personal character and work ethic to endure and thrive.

To my mom and dad, who loved me, fed me well, who painted a kickball field onto a public road in front of our house to promote outdoor play and exercise, and who paid for most of my undergraduate college education. Perhaps most important, my mom taught me persistence and my dad taught me kindness, both of which substantially impacted the writing of this book.

To my three sisters, who acted as three additional mothers when I was a kid. I am as mentally and physically healthy as I am as a fifty-plus-year-old male in large part because of the quantity and quality of the early life nurturing I received.

To my wife, Nhu, and daughters, Siana and Isla, who are the core inspiration for this book. I was never so aware of what I was eating until I met Nhu, and we'd like our daughters to continue to enjoy the same foods when they are adults that we do today. After a few decades on my own, immersed in scientific investigation, rock and roll music, and pick-up soccer, you gave me a second life with more dimensions and wonders than I could ever have imagined. I hope this book makes an impact on the way we all view food.

TABLE OF CONTENTS

PREFACE

Climate change is arguably the greatest challenge the human species will face in the coming decades. For quite some time, climate scientists have been warning about the chaos that will occur with an increase of just a few degrees in global temperature. News reports, internet articles, and blogs all seem to focus on the weather itself, the more severe and more persistent storms that will occur along with the fires, droughts, and floods. Less consideration appears to be given to what living conditions will be like in between the catastrophes and in regions that are not as directly or as severely impacted by changes in local weather conditions. If we break down life to its most basic parts, we eat, we drink, we sleep, and we do other things when we're not eating, drinking, or sleeping. This book focuses on the eating and drinking parts of daily life. How will the foods we eat and the drinks we imbibe change as the global climate changes? Food loss is not something that might happen in the future; we see the impact now in the rising costs of many natural foods. The scientific evidence is strong that food loss is already happening and will get much worse.

This is a reference book, if you will, a compilation of information about thirteen of the most endangered foods. It is more about plants and less about animals. However, there is a full chapter dedicated to fish, a relatively sizeable section about pollinating bees (honeybees in particular), frequent references to other insects, some props to gargantuan mammals that don't exist anymore, and discussion of human behavior (of course). There are cooking recipes in each chapter that are intended to better link a visceral experience to this serious problem (food is mood) and to provide instructions for simple and delicious ways to prepare and eat these foods while they are still available. There is also a list of the healthy nutrients, minerals, and phytochemicals contained within these endangered foods at the back end of the book, following the final chapter. The historical and scientific information was gathered from multiple sources including pubmed.gov (a search engine for peer-reviewed biomedical journal articles), federal agencies (including NOAA, USDA, FDA), Wikipedia, Encyclopedia Britannica, and some national and local news sources. The nutrients, minerals and phytochemicals listed at the back of the book were all sourced at the Linus Pauling Institute–Micronutrient Information Center. As a disclaimer, I donate a few dollars to Wikipedia on a yearly basis to keep the site running. All but one of the food images that open each of the food chapters were taken in our kitchen or dining room with an iPhone 10. All recipes were prepared and photographed in our kitchen.

I'm not a chef, a food critic, a dietician, or a food scientist. I am just a husband and a father that has developed a bit of an obsession with climate change. I'm also a college professor in the life sciences who studies how the brain controls animal behavior and who happens to have free access to a ton of peer-reviewed journal articles. I became concerned about global warming many years ago, when I was single, but it didn't produce the response in me then that it does now that I'm married and have young children. I call it an obsession with climate change, but not in a pathological way. I don't think and re-think the same global destruction scenarios one might envision if she/he were pathologically obsessed or producing some blockbuster planetary doomsday movie. I call it an obsession because the interest drove me to write this book, which was a bit of an undertaking. I felt this topic was worthy of a book because eating and drinking are a big deal and people might want to know this stuff.

We all eat at least a handful of the foods that are described. Some of us eat all of the foods that are described. This is not an issue for any singular demographic of modern-day society. This is an issue for all of us. Time to think and act, my fellow climate-change survivors.

CHAPTER 1

○ ○○○○○ ○○○○○○○ ○ ○○○○○ ○○○○○○○

Introduction

Food, glorious food! Never in the history of mankind has there been so much food on this planet, both in bulk and in variety. In just about any postindustrial nation, anyone can walk into a grocery store or supermarket and experience this on a near twenty-four-hour basis seven days a week. Dozens of varieties of fruits, vegetables, nuts and grains; whole wheat, rye, pumpernickel, sourdough, unleavened breads, and bread crumbs; hard and soft cheeses, creams, curds and Brie; red meats, white meats, various poultries, sausages, fish, and other seafoods. Including all of the different types of sauces, syrups, soft drinks, juices, candies, cookies, crackers, and other canned, jarred, packaged, and frozen foods, most supermarkets sell tens of thousands of different products. Some supermarkets sell 40,000 more products today than they did in 1990. There are roughly 38,000 grocery stores in the US creating a retail food industry that earns about $660 billion dollars per year. Total food sales in the US are about $1.4 trillion dollars every year, which amounts to approximately 5 percent of the total US economy (Food Marketing Institute, US Bureau of Labor Statistics).

Some of the foods we eat today have been with us for almost 10,000 years! To put this in perspective, plants have existed for about 700 million years while humans have been around for about 200,000 years. Well, the day of reckoning is upon us. The peak has been reached, and the food boom is already beginning to bust. In only fifty years or fewer, many of the delicious foods we eat will be things of the past that we describe to our kids, like rotary phones or the Dewey Decimal System. However, bulky phones with numbered dials causing the dialing of nine to take nine times longer than dialing one and ancient systems to organize books in libraries were replaced by better versions. This is not the case for food. Those foods that will disappear will not be replaced by better versions. Let's not lose these precious foods. Let's do the little things that will allow us to continue to wake up to a hot cup of coffee and end the day with some dark chocolate.

There are some great books out there that talk about how food production and industrial farming impact the environment and contribute to global warming (ex. *The Fate of Food* by Amanda Little; *Kiss the Ground* by Josh Tickell; *Food Fix* by Mark Hyman). This book takes the opposite perspective and focuses on the impact of global warming on the foods that we eat on a daily or weekly basis. The point of this book is simply to be a compilation of biographies for those endangered foods that have no voices to speak on their own behalf. Just as there is an American Society for the Prevention of Cruelty to Animals (ASPCA) and a World Wildlife Fund (WWF) that speak on behalf of abused and threatened domesticated and wild animals, perhaps there should be an American Association for the Prevention of Cruelty to Plants (the ASPCP), or even better, an American Association for the Prevention of Cruelty to Food (ASPCF), with this book highlighting the poster children.

Following a brief chapter on climate change and agriculture, each subsequent chapter describes an endangered food, from its known origin in the wild and symbolic or societal significance to its nutritional value, its earliest domestication, and the reasons why it will perish. Each food chapter finishes with two recipes, one for a standard dish and one for a nonstandard delicacy. The final chapter contains suggestions to help maintain the glory days of food, a list of little things that we can all do on a daily or weekly basis so that our children and grandchildren may grow up to directly experience fruit salads complete with apples and bananas, Mexican dinners with guacamole, and sporting events with beer and peanuts.

CHAPTER 2

○ ○○○○○ ○○○○○○○ ○ ○○○○○ ○○○○○○○

Impacts of Climate Change on Agriculture, Briefly

Global warming is occurring primarily due to the burning of fossil fuels. Carbon dioxide (CO_2) and other pollutants produced from fossil fuel combustion absorb infrared radiation from sunshine. In other words, as we burn more fossil fuels, the amount of CO_2 in the atmosphere increases, and more CO_2 stores more heat produced by the sun. This is analogous to a greenhouse trapping heat, which led to CO_2 being described as a "greenhouse" gas. The planet has undergone warming periods in the past, but never at such a pace as is occurring now.

How do we know CO_2 is increasing?

The Global Monitoring Division of the Earth System Research Laboratory (ESRL) was established about fifty years ago by the National Oceanographic and Atmospheric Administration (NOAA). This is a worldwide network of atmospheric monitoring stations. The Baseline Atmospheric Observatory in Mauna Loa, Hawaii, metaphorically the wise grandmother of this monitoring system, was established in the 1950s and has been continuously collecting atmospheric data since its inception, including CO_2 levels. Mauna Loa may be the perfect site for CO_2 measurements because of its remote location, undisturbed air, and distance from human activity. Measurements taken at the Baseline Atmospheric Observatory indicate that gaseous CO_2 levels have exceeded 400 parts per million (ppm, meaning that for every million floating molecules that comprise the atmosphere, 400 are CO_2). This is a level that has not been reached for about four and a half million years and continues to rise rapidly. A vast majority of climate scientists, like 97 percent or more, will say that we (humans) are responsible (Intergovernmental Panel on Climate Change, IPCC, 2013–2019). We did this. We can undo this.

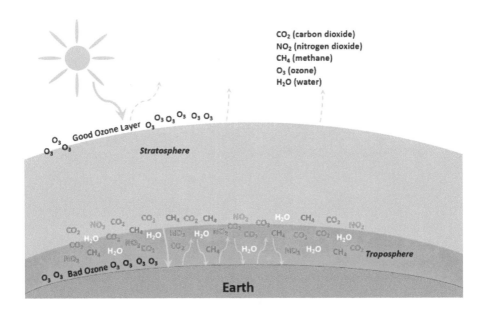

The illustration above depicts the greenhouse effect caused by carbon dioxide (CO_2), nitrogen dioxide (NO_2), methane (CH_4), and water (H_2O). Energy from the sun passes through the atmosphere (stratosphere and troposphere shown) and reaches earth, where it is reflected back toward outer space. Greenhouse gasses prevent some of this reflected energy from traveling back into outer space and trap heat near the earth's surface. Good ozone (O_3, not considered a greenhouse gas) in the stratosphere helps to keep the earth cooler by reflecting energy back into space. Bad ozone in the troposphere, at the earth's surface, is toxic to animals and plants.

How are plants impacted by elevated CO2 and other greenhouse gases?

I'm pretty sure we all remember from our high school botany lessons that plants "breathe" CO_2. So, we might put two and two together and think, somewhat correctly, that global warming and increased atmospheric CO_2 are good things for plants. They metabolize CO_2 on warm sunny days and, in the process, produce the oxygen that we breathe. One might guess that more CO_2 in the air should promote plant health. Indeed, this is the case. For instance, banana production has increased pretty steadily for the past five decades due to climate change.

Plants grow faster with higher levels of ambient CO_2. In fact, CO_2 canisters can be purchased commercially to aid in the growth of indoor plants. But excess CO_2 is only one factor that is raising the global temperature to dangerous levels. Nitrogen dioxide (NO_2) produced by high-temperature combustion of coal, oil, gas and diesel fuels is a major greenhouse gas produced mostly at the industrial level. Also, methane (CH_4) is released from decaying or burning organic materials or natural gas and is perhaps the most potent greenhouse gas. None of these chemicals will directly destroy plants, but they all raise the global temperature that then causes other effects that are severely deleterious to plant survival. Most of the dangers to our delicate food plants have been present for a long time, but have been manageable to a large extent. Continued global warming will make these situations unmanageable. Global warming is a bigger problem than just rising sea levels, bigger storms, and even greater food scarcity in developing world nations, as if that is not huge enough.

It is plausible to think that, as the planet warms, the longer growing seasons will also benefit plants. However, increased metabolism and a longer growing season produced by global warming are not necessarily compatible. All of the plants that exist today have evolved in a relatively cooler environment, with daily temperature, daily sunshine duration, and light intensity and wavelength fluctuating in harmony across the days and seasons for millions of years. Plants have developed complex genetic regulatory systems in direct response to these coordinated environmental changes. Think Darwin and reproduction of the fittest. Those plants that expressed the proper genes at the proper times with respect to daily and seasonal rhythms were more likely to reproduce and pass on their fitness genes to the next generation. These are the natural plants that exist today. They have thrived because their genes matched their environments.

While environment-driven changes in gene expression can occur on a relatively rapid timescale (elegantly described in Matt Ridley's *Nature via Nurture*) and may benefit some lucky plants and animals, in general genomes don't evolve as fast as the warming of the planet, and the overall increase in global temperature is causing climate zones to shift very quickly (John Daniel, NOAA; Susan Solomon, MIT). Many of these plants will not be able to adapt to such rapid changes in local climate. Some of the more endangered plants produce foods and beverages that have been beloved by humans across the globe for thousands of generations, and we are the generations that very likely will witness their extinction.

What are the environmental changes that doom these foods?

The primary factors that will drive food extinction are excessive ambient temperature, fresh-water scarcity, air pollution, extreme weather events, and heightened susceptibility to predators and disease. Some plants are very sensitive to temperature or require very specific temperature and humidity conditions during the growing season and across seasons to bear fruit. Apples, cherries, peaches, pears, and wine-making grapes fall into this category, some requiring 300 to 900 continuous hours anywhere between thirty-two and sixty degrees Fahrenheit every winter for proper bloom in the spring. Other plants use relatively large amounts of water to survive and bear fruit or beans. These climate-change victims include avocado, chickpea, and cacao plants that like it hot above ground and really wet below ground consistently for six to nine months. Finally, while elevated CO_2 may be beneficial to plant growth, the accumulation of other pollutants will produce negative impacts on plant health. Loosely speaking, when plants inhale ground-level ozone through their stomata, which are the rough equivalent to thousands of nostrils, they rust or oxidize from the inside out. In other words, just as oxygen strips electrons from iron to initiate rust (ferric oxide) formation and iron deterioration, ozone strips biological molecules of their electrons and important biophysical properties, leading to cell death.

As mentioned above, if unchecked, ground-level ozone will continue to rise during the next few decades. Also, methane is an important player here. Fossil fuel (natural gas) combustion, decaying food waste, thawing of permafrost wetlands, and maintenance of large cattle herds to support red meat consumption are major contributors to rising global methane levels. Methane can trap more heat than CO_2 by at least an order of magnitude (at least twenty times more). Bear in mind that as tropical rain forests and jungles continue to dry out and burn more and Arctic permafrost regions continue to thaw more, they will release more methane and produce more CO_2, rather than absorb or store it, further worsening an already bad situation.

A conversation about agriculture is not complete without at least one short paragraph about germs and insects. Increased global temperatures will increase the planet surface area that is habitable by most insects. Bugs will reproduce faster, and those once limited to the tropical or subtropical climates will begin to move more northward and southward of the equator and establish colonies in more temperate climates, in the corn and wheat fields, vegetable gardens, and orchards that we rely on for subsistence. These "prey" plants possess minimal immunity or defense against these critter invasions because they have not experienced them before. More or newly acquired bacteria, viruses, and bugs will tempt the use of more pesticides, which is wreaking havoc on pollinating bees and other insects critical to a balanced ecosystem (described a bit more in the Honey and Saving Endangered Foods chapters). In fact, it may not be possible to grow certain food plants outdoors without some sort of physical or chemical protection against predators and severe weather.

We are already actors in the horror story that is the US agriculture industry, created through forced use of genetically modified plants, application of human disease–causing herbicides, and twisting of property laws to put small farmers out of business. Clearly, genetically modifying food plants so that they can be treated with chemical herbicides or pesticides is not the answer. Perhaps genetically modifying food plants so they can better deal with climate change or withstand disease is a better idea. Genetically modified organisms (GMOs), including mutated plants, were developed decades ago. The agricultural industry is now turning to geneticists for assistance to a much greater extent given the enormity of climate-based food problems.

Okay, enough of the dreariness about the major causes of food extinction. The next few chapters highlight individual foods—thirteen delicious foods, to be exact—that would be screaming for help if they had mouths. Different foods on this list will suffer separate forms of demise. I describe these foods in alphabetical order. Why? Well, it's relatively arbitrary and lacks any subjectivity on my part (except for the choice of alphabetical order).

CHAPTER 3

o o o o o o o o o o o o o o o o o o o o o o o o o o o o o

The Non-Chilling Future of the Ubiquitous Apple

The apple has also served as a symbol of love, fertility, and abundance for about as long as it has been a food. The Greek god Dionysus offered apples to win the love of the goddess Aphrodite. Her Roman counterpart, Venus, has been painted holding an apple. Interpretations of the Bible and the forbidden fruit of the Garden of Eden have resulted in the apple being viewed as a symbol of knowledge, immortality, and temptation.

If you want to impress your elementary school teacher, you can give her or him an apple. To make spouses and children feel special, they are referred to as the "apple of one's eye." And there is nothing quite so American as apple pie. The apple is a card-carrying member of most fruit salads, it is bold enough to support a tart, it is delicious as a cold or hot cider, and it holds its ground on its own as well as any other raw fruit.

Can an apple a day really keep the doctor away?

Apples are high in fiber, vitamin C, other antioxidants, prebiotics (fructo-oligosaccharides, the fruit-based carbohydrates that feed the good bacteria in our guts), and simple sugars, but have a low glycemic index, roughly thirty to forty on the glycemic index scale. The glycemic index (GI) refers to the ability of a food to cause a rise in the level of blood glucose. The GI was developed in 1981 by Dr. Thomas Wolever and Dr. David Jenkins at the University of Toronto with the goal of identifying better food choices for people with adult-onset or type-2 diabetes (an insensitivity to insulin also referred to as "insulin resistance" in its early stages). The higher the GI, the more rapid and the greater the rise in blood glucose. Arbitrarily, glucose was assigned a glycemic index of 100, the highest value on the scale. All other foods are ranked relative to glucose. Foods with a GI of about seventy or higher cause large and rapid rises in blood glucose. These high-GI foods are not healthy because they promote systemic oxidative stress (think premature aging), lead to weight gain, which has its own set of unhealthy outcomes, and increase one's predisposition to adult-onset diabetes. So, replacing some high-GI foods with some apples should help keep the doctor away.

How old is the apple?

The sweet and juicy apples we know and eat today started off like the wild bitter crab apples many of us tried to eat as kids when unattended in our backyards or local parks. The wild Central Asian crab apple species, *Malus sieversii*, was the original dominant contributor to the genome of the cultivated apple, *Malus domestica*, and can be traced back to the Tian Shan forests of southern Kazakhstan some 4,000 years ago. However, bidirectional gene flow has resulted in the common domesticated apple being more closely related to the wild European crab apple, *Malus sylvestris*. There are currently more than 7,500 known cultivars (cultivated varieties) of apples bred for cooking, making cider or applesauce, or eating raw. Over ninety million tons of apples were produced in 2017, predominantly in China, and Americans eat more raw apples than any other fruit, except bananas (ten pounds of apples per year per person, twelve pounds of bananas per year per person) (United States Department of Agriculture, 2019).

Why are apples in danger?

For the last forty years, researchers in Japan have measured rainfall, time of bloom, time of harvest, and other environmental variables at experimental orchards in Nagano and Aomori. Effects of global warming are already being observed. Now these Japanese apple trees consistently mature more than a week earlier than they did forty years ago. They also bear fruit earlier in the season and produce apples that are softer and sweeter. This sounds like a good thing, but as global temperatures continue to rise, the impacts will be different.

Apples require a winter chill to end dormancy and enable buds to become flowers. Without this critical winter chill, leaf production can be delayed, blooming becomes irregular, fruit yield is reduced, and disease susceptibility is increased. Most apple trees need over 500 chill hours (at least twenty-one consecutive days of cold weather below forty-five degrees Fahrenheit or seven degrees Celsius) to remain healthy and bloom. This is in large part due to metabolism of a hormone known as ABA (abscisic acid) that inhibits apple tree germination. At warmer temperatures, ABA production and degradation rates are about equal, and the hormone maintains a constant level that is sufficient to maintain dormancy. When the temperature drops below the chill requirement, ABA degrades more quickly than it is produced and becomes undetectable by about three weeks after the initiation of the chill, ending dormancy and enabling apple blossom formation. As the planet warms, the number of chill hours will drop below critical levels, ABA levels will not decline sufficiently, and apple trees will not produce many apples and will eventually succumb to disease.

Japan is not the only nation experiencing apple growing problems. Right here in the US, apple production is already suffering. New York State is the second-largest producer of apples, with most grown in the Hudson Valley. Excessive late winter warming and severe early spring storms have devastated these apple orchards. Between 2012 and 2017, more than 2,000 farms closed, amounting to about 6 percent of all apple orchards in the region (New York Farm Bureau). Similar effects can be observed in the other apple-producing hotbeds across the nation, including California, Michigan, Pennsylvania, Virginia, and Washington. In addition to the severe weather, climate change is tipping the fruit-bug war in favor of the bugs. The three biggest killers of apples are mildew, aphids, and a fungus known as apple scab that attacks the leaves and fruits. Even the moderate-to-low-chill cultivars, apple trees that don't require as long of a cold spell, will suffer. There is one apple species that requires fewer than 100 chill hours. It is the Bahamian Dorsett Golden, which has only existed since the 1950s. Time to buy stock!

Apple trees will grow and produce fruit in most yards in the middle to upper latitudes of the United States. Saplings can be purchased at just about any garden center or home improvement store for about thirty to one hundred dollars. You will need at least two trees so that they can pollinate each other. I made the mistake of buying just one miniature apple tree and wondered why we weren't getting any apples. Then I ringed the trunk by accident when trying to prevent squirrels from climbing the delicate plant. The tree died, but an orphan emerged from the soil! Though it likely won't produce fruit, I'm tending to the orphan because it will still be nice to have a miniature apple tree in our front yard.

DELECTABLE APPLE DISHES

Fruit salad, applesauce, dried apple wedges, baked apples, apple and brussels sprout salad, apple cobbler, apple pie, apple fritters, apple tarts, apple crisps, candy apples, caramel apples

Standard Recipe: *All-American Apple Pie (Preparation time, 25 minutes; Cooking time, 50 minutes)*
    Ingredients:
    1/2 cup sugar
    1/2 cup packed brown sugar
    3 tablespoons all-purpose flour
    1 teaspoon ground cinnamon
    1/4 teaspoon ground nutmeg
    6 tart apples (ex. Granny Smith green apple)
    1 tablespoon freshly squeezed lemon juice (no seeds)
    Pastry for double-crust pie (9 inches)
    1 tablespoon butter
    The white from 1 large egg
    Additional raw sugar

Directions:
1. Preheat oven to 375° Fahrenheit
2. Peel, core, and dice the apples to roughly half-inch cubes
3. Mix together the flour, sugar, and spices in a small bowl and set aside.
4. Place the apple cubes in a larger bowl and toss the apples with lemon juice.

5. Add the flour/sugar/spice mixture and mix well to evenly coat all of the apple slices.
6. Line a 9-inch pie plate with the bottom crust and trim the edge to match the pie plate.
7. Fill the bottom crust with the seasoned apple mixture.
8. Cut the tablespoon of butter into small chunks and sprinkle evenly over the filling.
9. Roll remaining crust to fit top of pie and place over pie plate.
10. Trim the top crust to match the bottom, and seal and flute the entire edge.
11. Cut a few slits in top crust to allow the filling to breathe during cooking.
12. Beat the egg white into a foam and brush over the top crust.
13. Optionally, sprinkle some raw sugar on the top crust.
14. Cover edges loosely with foil and bake at 375° for 25 minutes.
15. Remove foil and bake until the top crust is golden brown (roughly 25 minutes longer).
16. Cool on a wire rack where kids and pets won't disturb it.

Nonstandard Recipe: Apple and Brussels Sprout Stir-fry (Preparation time, 10 minutes; Cooking time, 15 minutes)

Ingredients:

1 Granny Smith apple peeled and diced

½ cup of onion diced

1 tbsp garlic minced

1 ½ cups of mushrooms sliced

5 cups of water

3 cups of brussels sprouts trimmed and sliced in half

1 tbsp avocado oil

Salt and pepper

Directions:

1. Pour 5 cups of water along with a pinch of salt into a pot on medium to high heat.
2. Once water is boiling, pour the brussels sprouts and cook for about 4–5 mins to blanch them, then drain and transfer to cold water.
3. Pour the avocado oil into a pan on low heat, then throw in the minced garlic and diced onions and stir around for 4 minutes.
4. Increase the heat to medium and put in the drained brussels sprouts and sliced mushrooms.
5. Stir around for 3 minutes, then put in the apple pieces and cook for another 5 minutes.
6. Add salt and pepper to taste.
7. Remove from heat and serve on a platter.

Guacamole is to the Super Bowl as cranberry sauce is to Thanksgiving. Roughly eight million pounds of guacamole are consumed in a single day each year during the Super Bowl. Goodness! Ignoring the Super Bowl, consider the output of just one popular food business, Chipotle, which sells 97,000 pounds of avocados every day, or thirty-five million pounds per year! On average, Americans eat more than seven pounds of avocados per year, up from about two pounds per year in 2000 (Statista.com). We like to eat avocados, especially if someone else prepares them for us.

CHAPTER 4

o o o o o o o o o o o o o o o o o o o o o o o o o o o o

Science Portends the Pits for Avocados

Can an avocado a day keep the doctor away?

Avocados are often referred to as "alligator pears" and described as "buttery" or "nutty." Avocados are high in essential vitamin content, including biotin, folate (vitamin B9), riboflavin (vitamin B2, which cannot be stored in the body and must be consumed on a regular basis for optimal health), and vitamin K. Avocados also contain mono- and polyunsaturated fats (good fats) and a relatively higher level of copper (an important cofactor for enzymatic reactions in our bodies), along with other nutrients like thiamin, riboflavin and vitamin A, minerals (60 percent more potassium than bananas), and phytochemicals. Avocados have been shown to reduce dandruff, scabies (a contagious skin rash caused by a mite), and headaches.

How old is the avocado?

The undomesticated avocado (*Persea americana*) has existed for nearly 10,000 years, and prehistoric humans likely first ate wild avocados in what is now central Mexico. As is the case for most berries, it is likely that the avocado evolved to be eaten. The consumer would digest the flesh, but not the pit, which instead would be excreted in dung, acting as its primordial fertilizer. Nature is so smart! But wait, the avocado pit is huge. What animal can swallow and excrete an avocado pit? The answer is no animals that exist today are able to do this.

At the turn of the millennium, Dr. Connie Barlow coined the term "evolutionary anachronism" in her book *Ghosts of Evolution*. Evolutionary anachronism basically means that some characteristics of a species in existence today came about during coevolution with another species that has since become extinct. Evolutionary biologist Dan Janzen described the avocado as an example of evolutionary anachronism. Thousands of years ago, massive mammals (megafauna) like giant giraffe-sized ground sloths, car-sized armadillos, and colossal proto-elephants had no issues eating and excreting avocado pits. These behemoths all disappeared around 125,000 years ago at the end of the Pleistocene epoch, during the Ice Age. So, if the animals required for seed dispersal no longer existed, how is it that avocados persisted? It is believed that jaguars and squirrels contributed to avocado seed scattering, but most likely avocados were spared from extinction by human cultivation.

Domesticated avocado seeds have been found in the tombs of Incan mummies dating back to almost 3,000 years ago. However, the avocado was likely domesticated 4,000 to 6,000 years ago by the early Mayans. Though no remnants are present, the codices of the Mayans and Incans describe the avocado as a spiritual symbol and included it as the marker for the fourteenth month of the Maya calendar. Moreover, the classic Maya city of Pusilhá, in what is now southern Belize, is known as the "Kingdom of the Avocado," and the avocado tree is illustrated on the Maya ruler Pacal's sarcophagus at Palenque, Mexico.

Why are avocados endangered?

The avocado genus (*Persea*) has twelve species, and a majority of these species do not produce fruit. Avocado plants grow very large and require a lot of water to thrive. The avocado tree requires about 9 gallons of water for every ounce of plant, or 141 gallons (about the volume of two standard bathroom tubs) per pound. Relatively speaking, the water footprint of avocados is about ten times that of tomatoes, lettuce, or cucumbers. About 85 percent of the avocados eaten in the US are grown in Mexico and Chile. In Southern California, the number of avocado trees is steadily declining mostly due to drought and loss of fruit pickers. In Mexico and Chile, the number of avocado trees has increased, in part to make up for the loss in California. However, the number of trees that can be grown in these more tropical regions is limited by the availability of fresh water.

The number of avocado trees in existence use water faster than can be supplied by groundwater and river sources. So, this is not sustainable. Fortunately, a multinational team (Australia, Canada, Denmark, France, Mexico, Singapore, Spain, Sweden, and the US) of agriculturalists, geneticists, microbiologists, and statisticians just sequenced the entire avocado genome, which should help in identifying factors that will enhance avocado plant survival in the future. Sequencing and modifying the genomes of food plants may seem scary to many people, which is understandable. However, we use genetic screening and gene-based therapies in many ways to help humans already. Some forms of cancer can be successfully treated when the genetic makeup of the individual is known and the appropriate therapy is administered. This medical approach is now being applied to other disorders. Anyone can go to a website like Genesight.com to learn more about how genetic screening can better match drugs to patients with psychiatric conditions. Moreover, individuals suffering from single-gene metabolic diseases can now be cured with gene-based therapies. Gene editing in food plants is really no different. It presents a means to produce healthier and more robust plants, possibly saving them from extinction.

Try to grow your own avocado! Take an avocado pit and pierce it on three sides with toothpicks and then support the pit on a glass of water with the bottom third of the pit continuously submersed. Leave the glass and pit in a warm place that gets some sunlight. Be sure to keep the bottom of the pit wet with clean water. In a short time, the pit will crack, a root will grow downward into the water, and then a beautiful baby sprout will emerge upward! You might want to transfer the pit to soil when the root has substantially extended (prior to sprout development) to optimize growth.

DELECTABLE AVOCADO DISHES

Guacamole, avocado smoothie, avocado toast, avocado chips, avocado salsas and dips, avocado salads, avocado ranch dressing, frozen avocado pops, chocolate avocado pudding, sushi with avocado

Standard Recipe: Guacamole (Preparation time, 10 minutes; Cooking Time, 0 minutes)

Ingredients:

2 avocados

1 tbsp lime juice

¼ tsp salt

¼ tsp pepper

¼ tsp garlic powder

2 tsp diced red onion

2 tsp chopped fresh cilantro

or 1/2 tsp dried cilantro

Directions:

1. Cut the avocados in half and use a spoon to remove the pits.
2. Scoop the avocado flesh from the skins into a medium-sized bowl.
3. Mash the avocado flesh with a fork and add the rest of the ingredients.
4. Mix well and add more seasonings to your personal taste.
5. Serve immediately.

Nonstandard Recipe: Avocado smoothie (Preparation time, 10 minutes; Cooking Time, 0 minutes)

Ingredients:

2 avocados

4 cups of plant-based milk

4 tbsp of sugar

Directions:

1. Cut the avocado in half and remove the pit.
2. Scoop the avocado flesh into a blender.
3. Pour the plant-based milk and scoop the sugar into the blender.
4. Blend well and enjoy!

Banana "trees" contain no woody material. As such, banana plants are herbaceous, in fact the largest herbaceous flowering plants on earth, and are not truly trees. Humans don't know how to peel bananas. Just watch a monkey in a zoo. They don't pull on the stem; they open the fruit from the blackened end, where it is soft and can be penetrated easily with fingernails. The banana plant (Musa genus) is tropical and currently grows in Africa and the Americas, along with Southeast and Southern Asian, Melanesia, and the Pacific Islands. There are at least fifty species of bananas in the Musa genus (which also includes plantains) and hundreds of cultivars characterized by peel color, peel thickness, taste, fruit size, and resistance to disease.

CHAPTER 5

○ ○○○○○○ ○○○○○○○ ○ ○○○○○ ○○○○○○○

Did You Know Bananas Are Herbs?

**Bananas are often touted as the
perfect after-workout food**

After longer-duration physical exertion, eating a banana replenishes lost water, carbohydrates, and potassium, along with vitamin B6, vitamin C, copper, magnesium, and anthocyanins, which are plant pigments that reduce oxygen radicals. Eat a greener banana and we're ingesting more starches or polysaccharides, which are long-chain carbohydrates that serve as prebiotics (through colonic fermentation) but mostly pass through the digestive system and have a low glycemic index (GI) of about thirty. Eat over-ripened bananas and we're eating more simple sugars, doubling the GI to 60. Remember from Chapter 3 that foods with GIs of seventy or higher are not considered to be healthy to eat, so eating over-ripened bananas is pushing the health limit with respect to GI.

Regardless of ripeness, eat a banana and we're eating radiation. Yes, bananas are radioactive due to the presence of the isotope potassium-40, which decays to argon and gives off a neutrino and a gamma ray. We would have to eat 100 bananas in a day to reach the average daily exposure to radiation most humans experience, so don't walk away thinking bananas cause cancer. There is no evidence for that. In fact, because they contain antioxidants like vitamin C and anthocyanins, bananas may reduce one's predisposition to cancer.

Where did bananas come from?

Banana cultivation can be traced back at least 9,000 years to the Kuk Swamp in the Western Highlands of Papua, New Guinea. Traders from New Guinea introduced bananas to India, Africa, and Polynesia, and then, about 2,400 years ago, Alexander the Great invaded India and brought bananas back to Europe. It wasn't until the late 1800s that bananas were first eaten in the US. In 1876, bananas (then, the Gros Michel cultivar) were displayed on a forty-acre field at the Centennial Exposition in Philadelphia. Before the turn of the century, Andrew Preston and Minor Cooper Keith began importing bananas from Central America and the West Indies and selling them through their Boston Fruit Company, which then became the United Fruit Company, and then Chiquita.

When hearing the words "Chiquita banana," I believe most people about my age remember Carmen Miranda, the film star also known as the Chiquita Banana Lady with the catchy jingle from the TV commercials. The song itself was actually born in the 1940s and Ms. Miranda first made it popular in cinema commercials prior to TV. But, before Ms. Miranda began gleefully singing about bananas, operators of the United Fruit Company, masquerading as philanthropists, were wreaking havoc on Latin American banana workers and their communities, forcing horrific working conditions, preventing the formation of unions, and intimidating national governments, all to minimize cost to American consumers and maximize profits. This appalling corporate greed culminated on December 6th, 1928, in Ciénaga, Columbia, during the Banana Massacre where as many as 3,000 Columbian people, workers and their families labeled as rebels by the Columbian government, were machine gunned down by the Columbian troops commanded by General Cortéz Vargas. Other members of the "Banana Republics," including Costa Rica and Guatemala were similarly manipulated and oppressed by the United Fruit Company which resulted in social and political instability that lasted for most of the twentieth century.

The domesticated banana that is a brilliant yellow and familiar to most of us in the West is the Cavendish. The Cavendish is grown primarily in South America where it takes at least nine months to produce fruit. All edible bananas available today have been hybridized from *Musa acuminata*. However, genetics work against reproduction of edible banana plants by seeds because the plant is triploid. In other words, it has three sets of chromosomes and rarely produces viable seeds. Humans are diploid and have twenty-three pairs of chromosomes, so the math is easy. In the production of human eggs and sperm cells, the twenty three pairs of chromosomes are split so each egg or sperm cell has only one half of the twenty-three pairs. Shortly after conception, the twenty-three egg chromosomes and the twenty-three sperm chromosomes come together to form twenty-three new chromosome pairs in the cells of the offspring, just as in the parent.

Having three sets of chromosomes complicates things for the banana because eggs and sperm cells tend to have the wrong number of chromosomes, and so sexual reproduction produces inviable offspring. Instead, when the mature plant dies, small protuberances (side shoots or suckers) near the roots can be picked and replanted. This means that all of the Cavendish bananas we buy in the produce section are not just siblings; they're all clones! In fact, every Cavendish plant can be considered part of one extremely large collective organism. This is great for consistency in flavor, but is not so good when the plant is challenged by disease or parasites. In fact, the Cavendish is not the original American favorite. As mentioned in a previous paragraph, just a few generations ago, Americans ate the Gros Michel, or "Big Mike" banana, but it was nearly completely decimated in the 1950s by a root-eating fungus known as Panama disease (described in more detail below).

What is the biggest climate-change threat for the banana?

As mentioned in the opener, warmer temperatures and higher levels of atmospheric CO_2 can benefit plants. Bananas are a prime example. A recent study by Dr. Dan Bebber and Dr. Varun Varma (both at University of Exeter in Exeter, England) revealed that over the past fifty years, banana cultivation has increased in the twenty-seven countries that provide nearly 90 percent of the bananas that we eat. This was due in part to climate-change factors that improved growing conditions and accelerated plant and fruit maturation. The good news is that some of these benefits will continue for a little while for banana growers in Ecuador and some parts of Africa. However, the largest banana-producing and consuming nation, India (producing mostly miniature Cavendish and Poovan bananas, but also Bluggoe and Nendran cultivars), and the fourth leading producer in the world, Brazil, along with numerous other countries (Columbia, Costa Rica, Guatemala, Honduras, Panama, and the Philippines), will suffer. By 2050, any positive effects of climate change will be substantially lessened, and the output from the endangered regions will be reduced by more than 80 percent.

In addition to the negative impacts of increasing temperature and extreme weather, climate change is allowing a banana-killing fungus that causes "Fusarium wilt," also known as Panama disease (Tropical Race 4, TR4), to run rampant through Latin America, and it is now present in Australia. This is highly problematic because this fungus originated in Southeast Asia, and so Latin American and Australian banana cultivars have minimal immune defense to these killers. Also, while once nearly completely eradicated in Australia, black sigatoka, which causes black leaf streak disease in bananas, is again on the rise in Australia and Asia.

Scientists from at least nine countries (not including the US) are working together to understand how drought-resistant bananas are able to survive the stress of reduced water availability. Those banana varieties that best deal with reduced hydration are able to utilize their genomes to the max. They hyper-express stress-response genes that allow them to minimize accumulation of damaging oxygen radicals, and maintain higher water content, normal water/salt balance, and better respiration. Knowing how successful banana plants deal with stress may allow for creation of new varieties that will withstand climate change and may also benefit the survival of other endangered food plant species.

Some effort is apparent on the industry level, too. One of the biggest sellers of bananas in the US, Dole, has shown that it can operate banana plantations in Costa Rica with 80 percent less water and reuse the plastic covers that protect the plants. While they've shown proof of principle that better agricultural practices are possible, this was just one of many commercial plantations in Latin America that would need to be converted for this effort to be meaningful, and there is little evidence of progress in this direction. Instead, banana growers seem to think that they will escape climate change by simply growing bananas in places that are more welcoming. There appears to be less commercial effort to protect bananas from a climate-change disaster compared to the other endangered foods described in this book.

DELECTABLE BANANA DISHES

Banana bread, banana split ice cream sundae, dried banana chips, deep fried bananas, banana pancakes

Standard Recipe: Banana bread (Preparation time, 20 minutes; Cooking Time, 60–75 minutes)

Ingredients:

4 ripe bananas

1 tsp ground cinnamon

1 tsp baking powder

2 cups of all-purpose flour

1 tsp baking soda

2 large eggs

½ tsp salt

8 tbsp (1 stick) of room temperature butter

½ tsp vanilla extract (optional)

½ cup of coconut sugar (optional)

Directions:

1. Preheat oven to 350° Fahrenheit.
2. Combine the bananas, coconut sugar, and butter in a mixer and mix well.
3. Add the eggs one by one into the mixer followed by the vanilla extract (optional).
4. In a small bowl, combine the dry ingredients: flour, ground cinnamon, baking powder, baking
5. soda, and salt.
6. Slowly add the dry ingredients into the mixture and mix until the flour disappears.
7. Pour the batter into a prepared buttered loaf pan.

8. Bake 60–75 minutes or until a toothpick inserted into the center of the pan comes out clean.
9. Cool on a rack for 10–15 minutes, then remove bread from the pan.
10. Let the bread cool for another 5–10 minutes before slicing, and enjoy!

Nonstandard Recipe: Banana chocolate chip cookies (Preparation time, 15–20 minutes; Cooking Time, 12–15 minutes)

Ingredients:
16 tbsp (2 sticks) of room temperature butter
½ cup of brown sugar or coconut sugar
1 ripe banana
1 egg
½ tsp vanilla extract
2 cups of all-purpose flour
1 tsp baking powder
1 tsp baking soda
½ tsp salt
½ tsp ground cinnamon
1 cup of semi-sweet chocolate chips

Directions:
1. Preheat oven to 350° Fahrenheit.
2. Line baking sheets with parchment paper.
3. Combine the butter, coconut sugar, banana, egg, and vanilla extract in a mixer and mix well.
4. In a small bowl, combine the dry ingredients: flour, baking powder, baking soda, and salt; and stir.

5. Slowly add the dry ingredients into the mixer and mix until the flour disappears.
6. Slowly add in the chocolate chips.
7. Scoop the dough onto the baking sheet.
8. Bake at 350° for 12–15 minutes or until the edges are lightly browned.
9. Remove from oven and transfer cookies into a rack to cool completely.

Beer is the most commonly imbibed beverage that contains alcohol. Once demonized during the prohibition era (from 1920 to 1933), alcohol has rebounded as THE socially acceptable vice. A Gallup poll found that over 40 percent of Americans prefer beer to other alcohol-containing beverages. When the economy slows, people buy more beer to drown their sorrows. When the economy flourishes, people buy more beer to celebrate. From our first US president's home (Mount Vernon supported a brewery), to a high school student's first drink, to a Supreme Court justice's (the Honorable Brett Michael Kavanaugh) favorite pastime, beer has a special place in American culture.

CHAPTER 6

○ ○○○○○○ ○○○○○○○ ○ ○○○○○○ ○○○○○○○

Simply, Beer

Prior to the raising of the legal drinking age from eighteen to twenty-one years of age in the 1980s, college kids would buy kegs of beer and lug them up to their dormitory rooms for Thursday or Friday night "keggers." Across the nation, Thursday night remains the beginning of the weekend for upper-level college students with a higher priority for social engagement, resulting in avoidance of Friday morning lectures (by not scheduling or not attending them). Also in the 1980s, President Jimmy Carter signed Bill H.R. 1337, which negated any tax on beer brewed and consumed at home, and in the late 1970s, Charles "Charlie" Papazian, a nuclear engineer by profession, founded the first Association of Brewers and the American Homebrewers Association, which is now a part of the

Brewers Association. The Brewers Association has grown to include over 7,000 brewers, representing over seventy types of beers (both ales and lagers), who are totally consumed with the craft of brewing beer. My PhD thesis advisor gave beer brewing a try and won an award at a local brewing competition in Charlottesville, Virginia. There is nothing odd about emptying three or four sixteen-ounce cups of beer at our favorite sporting events or picking up a couple of forty-ounce bottles, the Winnebago of beers, on our way to a house party. I worked for Budweiser delivering beer when I was in college. After a hot summer day full of manual labor, there was no better relief than an ice-cold Budweiser. It was light, tasty, easy to drink, it replenished lost carbohydrates (lugging beer is hard work!) and was perfectly refreshing. We love beer!

Who can I thank for creating beer?

Chemical tests from ancient pottery discovered in what is now Iran (then, part of Sumer) date the earliest production of beer to about 7,000 years ago. Since cereal grains were unavoidably exposed to wild airborne yeasts, there was a propensity for natural fermenting and beer drinking to arise soon after the development of cereals for human consumption. Archaeological evidence indicates beer production on a domestic level in Neolithic Europe about 5,000 years ago. Given the fact that beer recipes were included in the writings on ancient Egyptian papyrus scrolls dated 5,000 years ago and in poems by the ancient Sumerians nearly 4,000 years ago, it is apparent that beer has had considerable cultural significance for a very long time. In fact, the Egyptians even worshipped a goddess of beer called Ninkasi, and recited a hymn to Ninkasi that aided in passing the recipe for beer from generation to generation.

German monks first added hops to beer recipes near the end of the twelfth century due to its attractive flavor and natural preservative properties. Soon after, hops became popular as a beer ingredient outside of the monastery. German and Belgian monasteries are still considered some of the best beer brewers on planet Earth. Westvleteren in Belgium produces three beers with international reputation. These beers are sold in amber bottles without labels. They are identified by the color of the bottle cap, and they are the only Trappist beers (meaning of the Catholic religious order the Order of Cistercians of the Strict Observance) that don't have the Trappist logo on the bottle. I drank my first Trappist beer in Belgium at sixteen years of age during a tour with my town club soccer team. I remember that it was heavy and bold. During the trip, we visited pubs on a nightly basis, and I was taught that drinking some beers at very cold temperatures is not the best way to experience the intended flavors and that some beers contain a lot of yeast that settles to the bottom of the bottle, so drinking no more than 70 to 80 percent of these beers is recommended to avoid the yeast flavor (the yeast is not unhealthy to consume).

There are twenty-four Benedictine brewers in German monasteries, with half of these monasteries located in Bavaria. Eight abbey breweries on the Kruezberg Mountain in Franconia recently formed an alliance largely to maintain their brewing tradition. The chairman of this association, Father Lukas Wirth from the Benedictine Abbey of Scheyern, has proclaimed, "We are bound not only by the historical and cultural importance of the individual breweries, but above all by our shared values"— principally, to support social causes. Somewhat off-topic here, in 1759, Arthur Guinness signed a 9,000-year lease for a lot in Dublin (eventually purchasing the property outright) that remains the St. James Gate Brewery and the hub of Guinness beer production to this day. Talk about confidence in your product! Your buddy Joe may think he is the original beer enthusiast, but Joe has a lot of competition for that honor dating back thousands of years.

Industrial-level production of beer is required because so many people drink beer. Beer started to be produced for the masses during the industrial revolution and was aided by production of affordable hygrometers (that measure humidity) and thermometers enabling better control over fermenting/brewing conditions and standardization of recipes. Given its mass consumption, one might think more would be known about the health impacts of beer. Beer gets a bad name in association with the "beer belly," though it is probably not the beer that is responsible for the accumulation of belly fat. Most likely, it's the processed carbohydrate-dense foods that accompany the beer and maybe the lethargy that follows. Moderate consumption of beer can increase good fats in our bloodstreams, and the taste of beer can provide a pleasure sensation. However, there are healthier ways to regulate blood lipid content, and the obvious element of beer that produces a psychological impact is the alcohol. At low amounts, alcohol can facilitate the actions of inhibitory neurotransmitters in the brain, helping one to feel more relaxed. At higher doses, alcohol drills holes into brain cells, making them leaky and dysfunctional. Everyone is better at pool or darts at the local pub after a couple of beers. However, by the third or fourth beer, we typically suck. This is the biphasic impact of alcohol on brain function, attention, and fine motor skills.

Even hearty beers are no challenge for climate change

Beer brewing is a rather delicate process. The ingredients, including barley, hops, yeast, and water, need to be in perfect balance to produce a full, tasty beer. At another level, these beer constituents need to be healthy so that the ingredients within the ingredients (starches, proteins, enzymes, etc.) are in harmony to optimize fermentation. Perhaps the biggest issue for the future of beer production is the barley grain that is required for fermentation. More frequent, persistent, and severe droughts will drive down barley yield almost entirely. Those barley plants that actually survive will be metabolically altered by climate change and will either produce grain with too little starch or too little enzyme to process the starch properly to feed the yeast. On top of this, barley farmers will be driven more to sell their barley as livestock feed rather than as beer substrate if red meat consumption does not decline. Barley is certainly a healthier option to feed to cattle than corn or soy meal (healthier for the cattle and the beef consumer).

The largest beer brewing conglomerates on the planet are now teaming up with geneticists to try to save barley. For instance, Anheuser-Busch InBev is working with a company called Benson Hill Biosystems to apply machine learning–assisted genomics to the derivation of new barley strains that will thrive on a warmer, more volatile planet. AB InBev is supplying terabytes of genetic, environmental, and growth data on barley strains it has collected over decades of research. Benson Hill Biosystems is applying state-of-the-art computer algorithms to these data to determine which strain crosses are most likely to produce barley plants with desired characteristics. The machine-learning algorithms permit rapid analysis of millions of different genetic combinations on a background of varying growing conditions, permitting relationships between genes and environment to be understood at light speed relative to the time needed for more traditional approaches, like physically crossing different barley varieties and waiting for them to mature under different environmental conditions. Deschutes Brewery in Bend, Oregon, and Carlsberg Brewery in Copenhagen, Denmark, are taking similar approaches to better understand and refine their brewing processes.

For the sake of beer, let's keep our fingers crossed that machine-learning algorithms are all they are touted to be. If you want to design your own beer via artificial intelligence, go to IntelligentX. com, input some desired characteristics, and let the machine-learning algorithm do the number crunching to create YOUR beer. Ten cans will be sent to you to enjoy at your leisure (and cost).

DELECTABLE BEER DISHES

Straight from the bottle, beer-battered chicken, beer-infused chili, beer cheese

Standard Recipe: Enjoying straight from the bottle (Preparation time, 5 seconds; Cooking Time, 0 minutes)
Ingredients:
1 bottle or can of beer

Directions:
1.   Open bottle with hand or bottle opener

Nonstandard Recipe: Beer-battered Cod (Preparation time, 30–40 minutes; Cooking Time, 5–10 minutes)
Ingredients:
Frying oil (canola/vegetable)
2 lbs of cod fillets cut into 1-inch strips
One 12-oz bottle of beer
3 cups of all-purpose flour
1 cup of salt
Ground black pepper
1 cup of garlic powder
½ cup of paprika
5 tbsp of cayenne pepper (optional)

Directions:
1.   Pour the oil into a dutch oven or pot and heat to 375° Fahrenheit.

2. Combine the dry ingredients into a small bowl: flour, salt, garlic powder, paprika, and cayenne pepper (optional).
3. Pour the bottle of beer into a large bowl.
4. Slowly add the dry ingredients into the large bowl of beer and stir to mix well.
5. Pat the cut-up fish strips dry and sprinkle salt and pepper on both sides.
6. Dredge the fish strips into the beer batter, and carefully slide the strips into the hot oil.
7. Cook fish strips for 5 minutes and turn over midway.

Cherry blossom festivals are held in Washington DC as well as Washington State, and in states in between such as Michigan, California, Oregon, and Wisconsin. Millions of people travel thousands of miles to view the early spring cherry blossoms. My wife and I attended the cherry blossom celebrations in Washington DC a few years back. The parade was extravagant, but the trees were mesmerizing. Row after row of blooming cherry trees, each an individual corsage of densely packed, small, pink flowers decorating the walkways and the riverside.

CHAPTER 7

○ ○○○○○○ ○○○○○○○ ○ ○○○○○ ○○○○○○○

Cherries Fueled Armies and Libidos

Cherries are symbolic

In Japan, the cherry blossoms, *sakura*, signify the brilliance and brevity of life since the cherry blossom flowers peak in near unison at about two weeks after the start of blooming and then quickly fall to the ground. As they bloom, hundreds of thousands of these delicate trees capture the attention of the nation, and they celebrate the moment with vigor in a thousand-year-old custom called *hanami*, or "watching blossoms." In 2018, climate change caused Japanese cherry trees to bloom in October!?! Horticulturalists like Hiroyuki Wada at the Flower Association of Japan surmised that large amounts of leaf loss resulting from a strong typhoon caused bloom-inhibiting hormones produced in leaves to be decreased, prompting this seasonal faux pas. If you want to read a captivating description of the history of cherry trees in Japan, check out *Sakura Obsession* by Naoko Abe.

On a more intimate level, cherries have been associated historically with ripeness of a young woman's sexuality, and the ability to tie a cherry stem into a knot with one's tongue and teeth serves as a dubious measure of sexual prowess. The cherry bomb is the literal bomb of firecrackers, complete with a green stem (fuse) and spherical red bomb composed of an explosive core, paper, and a fibrous outer shell (now illegal, so don't go try to buy one). Worldwide, there are probably few children who have not savored a Maraschino cherry gently placed like a wedding ring on a whipped-cream pillow atop an ice cream sundae or chocolate cupcake. My grandma and grandpa had what I believe was a Montmorency cherry tree at the furthest extent of their backyard in Wethersfield, the oldest town in Connecticut. This tree was huge (from a ten-year-old's perspective), big enough to climb two or three branches upward, and produced sour cherries that we would eat straight from the twig.

Cherries have a checkered past

Based on fossilized cherry pits found in prehistoric caves, cherries (the genus, *Prunus*) are believed to have originated anywhere from 5,000 to 7,000 years ago in the region between the Black and Caspian Seas at the borders of what are now Eastern Europe, Asia, and the Middle East. The first known writings describing cherries come from ancient Greece. About 2,300 years ago, Diphilus of Siphnos, an ancient Greek physician, touted the usefulness of cherries as a diuretic, and philosopher Theophrastus included cherries in his catalog of fruits entitled *History of Plants*. Roughly 200 years later, cherries were introduced to Europe by Lucius Licinius Lucullus (say that name three times fast), an influential Roman general and politician. Roman soldiers carried cherries as part of their rations, and the pits that were discarded along marching routes became a bulk of the cherry trees that grew throughout the empire. Viewed from 5,000 feet above, the pattern of cherry tree groves maps out the marching routes of Roman soldiers.

Black cherries (*Prunus serotina*) and chokecherries (*Prunus virginiana*) are native to North America, but the cherries we eat the most, Bing cherries (*Prunus avium*), were imported from Europe in the 1600s. These cherries were named for their first American cultivator, Ah Bing, a Chinese immigrant who toiled relentlessly on the farm of Seth Lewelling at the turn of the twentieth century to establish orchards in the American West, only to be denied re-entry to the US by the Chinese Exclusion Act after a visit to his homeland.

Cherries contain high levels of prebiotic fiber, antioxidants (vitamin C, polyphenols), anti-inflammatories (beta-carotene), vitamin K, potassium, copper, and manganese.

Why is temperature so important for spring cherry blossoms?

Similar to the apple tree, cherry trees require a winter chill period and a dormancy that outlasts the final spring cold spells. Warmer weather in late winter can prompt cherry trees to leave dormancy and bloom. They then become susceptible to catastrophic damage if there is a subsequent frost. This is occurring more and more frequently. In 2002, there was a total loss of cherry crops in Michigan due to a warm spell followed by a cold snap. Similarly, in 2012, another warm period followed by severe cold and wind in Michigan destroyed nearly all cherry crops. Nearly 185 million pounds of cherries were lost. Local farmers downplayed the tragedy and described the event as a "once in a generation occurrence." Then, again, in 2015, about one-third of the Michigan cherry yield was lost. Maybe the three years in between these misfortunes seemed like a generation to Michigan cherry farmers.

In Washington State, similar warm-cold events have hindered cherry tree growth. But in the Great Northwest, another factor impacting cherry yield is severe heat during the summer. Cherries don't like to get too hot. They stop growing, they get soft, and they tend to separate from their stems, which impedes harvesting. Also, it is unhealthy for cherry pickers to be gathering fruit in 100-degree-Fahrenheit weather. Where there used to be nine hours to pick before the day got too hot, now there are only six hours.

Who will save cherries from extinction?

Without outside assistance, more severe and unpredictable late winter weather and greater-intensity summer heat will destroy nearly all cherry plants. As in other regions where cherries are grown, in Washington DC the local microclimate is hotter and drier, putting more stress on cherry trees. An additional threat to cherry trees in our nation's capital is the rising water level in the Potomac River that overflows the tidal basin wall near the Thomas Jefferson and George Mason memorials. For the last decade, twice daily, walkways near the cherry tree groves are flooded, and this water reaches the roots of the cherry trees. This water has a higher salinity and degrades the roots of cherry trees. In the spring of 2019, the Trust for the National Mall and the National Trust for Historic Preservation in collaboration with the National Parks Service enacted the "Save the Tidal Basin" campaign. This campaign helps fund a National Mall Tidal Basin Ideas Lab that is working to restructure the tidal basin walls (minimally modified since 1882) and address other Tidal Basin challenges. The American Express Foundation has donated $750,000 to this project.

In Japan, outside assistance is coming in part from bear behavior. Animals also need to escape the increased global temperatures and are doing so by migrating toward the planetary poles or up mountainsides. The Asiatic black bear is no exception and is migrating to higher altitudes. These bears eat cherries at lower altitudes and then defecate the seeds at higher altitudes. In this way, cherry trees have followed bears up the mountainside to cooler microclimates. Will this be enough to save cherries from annihilation?

DELECTABLE CHERRY DISHES

Cherry cobbler, cherry pie, cherry tarts, dried cherries, granola cherry bars, cherry sauce

Standard Recipe: Cherry Oatmeal Crisp
(Preparation time, 20–30 minutes; Cooking Time, 30–35 minutes)

Ingredients:

6 cups pitted cherries

1 ½ cups of brown sugar

2 tbsp cornstarch or tapioca starch

1 tsp of salt

Juice of half a lemon

¾ cup all-purpose flour or plant-based flour

1 cup of old fashioned rolled oats

1 tsp ground cinnamon

½ cup of unsalted butter softened

Directions:

1. Preheat the oven to 400° Fahrenheit.
2. Butter a 9-inch baking dish.
3. In a bowl, combine the cherries, half of the sugar, corn or tapioca starch, half of the salt, and lemon.
4. Mix well and be sure to coat the cherries.
5. In another bowl, mix the flour, rolled oats, ground cinnamon, butter, and the rest of the sugar with your hands into coarse meal.
6. Pour the cherry mixture into the baking dish and sprinkle the dry ingredients evenly on top.

7. Bake at 400° 30–35 minutes until the topping is golden.
8. Remove from oven and let it sit for 10 minutes before serving.

Nonstandard Recipe: Cherry Cookies (Preparation time, 10–15 minutes; Cooking Time, 12–15 minutes)

Ingredients:

16 tbsp (2 sticks) of room temperature butter

¾ cup of coconut sugar

2 eggs

½ tsp vanilla extract

2 cups of all-purpose flour

1 tsp baking powder

1 tsp baking soda

½ tsp salt

1 cup of dried cherries

Directions:

1. Preheat oven to 350° Fahrenheit.
2. Line baking sheets with parchment paper.
3. Combine the butter, coconut sugar, egg, and vanilla extract in a mixer and mix well.
4. In a small bowl, combine the dry ingredients: flour, baking powder, baking soda, and salt and stir.
5. Slowly add in the dry ingredients into the mixer and mix until the flour disappears.
6. Slowly add in the dried cherries.
7. Scoop the dough onto the baking sheet.

8. Bake at 350° 12–15 minutes or until the edges are lightly browned.

9. Remove from oven and transfer cookies into a rack to cool completely.

If you like guacamole, you probably like hummus too, which is made from the legumes known as chickpeas (Cicer arietinum), or garbanzo beans. And if you like chickpeas, you're not alone. Chickpea consumption is on the rise globally in large part due to increased incorporation into healthy snacks and the use of chickpea flour, or gram flour, in baked goods. Chickpea flour is also known as besan in South Asia and is a preferred form of starch in Bangladesh, India, Nepal, and Pakistan, in part because chickpea flour retains nutrients and fiber content better than other refined flours. About 90 percent of the world's chickpeas are produced in South Asia, 65 percent in India.

CHAPTER 8

o oooooo oooooooo o ooooo oooooooo

Chickpeas, the Thirstiest Endangered Food

This bean is a critical source of nutrition for hundreds of millions of people in developing nations and the major source of protein for 20 percent of the world population. Chickpeas are also high in folate, beta carotene (think carrots), calcium, copper, iron, magnesium, manganese, phosphorus, and zinc. The high protein and fiber content of chickpeas make them harder to digest, so they also act to reduce appetite by remaining in the digestive system longer.

This ancient legume is blessed with two funny names

Chickpeas are members of a legume group or "pulse" where the seeds grow within pods. This pulse also includes lentils and dry peas. The word *chickpea* comes from the French word *chiche,* which was adapted from the Latin phrase *cicer arietinum,* or "small ram," reflecting the shape of this bean. The word *garbanzo* has no definitive origin; some say Basque, *garau;* some say Greek, *erébinthos;* some say Old Spanish, *arvanco,* which became modified to *garvance* upon introduction to English and was then further modified to *calavance,* which is generic for "bean." Chickpeas are also referred to as Bengal gram, Egyptian pea, chana, or chole.

The first human consumption of chickpeas is believed to have originated in southern Turkey over 7,500 years ago (possibly 11,000 years ago). Remains have been discovered there in the aceramic levels of archaeological digs in Jericho and Çayönü. So, chickpea consumption preceded pottery use! Chickpeas are now grown in Asia, Australia, Canada, Europe, Mexico, South America, North Africa, the Middle East, the United States (California, Idaho, Montana, North Dakota, and Washington), and the state of Maharashtra, India. Two-thirds of the chickpeas eaten in the US are grown locally. Chickpeas also come in red, green, and black varieties. A rare variety, the black chickpea (*ceci neri*), is grown in Apulia, Italy.

The chickpea is not ram-tough

While it sounds like there is large degree of variety in chickpeas, domestication has greatly reduced genetic diversity in cultivated strains (Clarice J. Coyne, USDA). Upon domestication, chickpea plants were selected and bred for the main characteristic of keeping the seed on the plant, which greatly facilitated cultivation. Unfortunately, this selection process has also made cultivated chickpea plants less adaptive and more sensitive to climate change.

If you thought the avocado required a lot of water to produce fruit, consider the chickpea plant, which requires eight times more water per ounce than the avocado and consistently moist soil to survive and generate pods. Not only that, but chickpeas have a very long growing season, so maintaining consistently moist soil for such a long period of time is difficult and expensive. Conservative estimates place present-day chickpea production at half of what it was in the twentieth century due to worldwide drought. As we say goodbye to the avocado, for the same reasons, we also say goodbye to the chickpea. Well, maybe not.

A hybrid genetics–agriculture approach might preserve the chickpea

The genome of the chickpea has been sequenced, and genes that may provide resistance to heat, drought, insects, and disease are beginning to be identified. Agricultural scientists like Shiv Agrawal at the International Center for Agricultural Research in the Dry Areas (ICARDA.org) in Lebanon are now turning to wild species and landraces to find combinations of chickpea genes that will permit these plants to continue to grow on a less habitable planet. Also, Dr. Rajeev Varshney and his team at the International Crops and Research Institute for the Semi-Arid Tropics (ICRISAT) in India have identified genes that will hopefully help chickpeas survive in much drier and hotter microclimates.

Likewise, Dr. Eric Bishop von Wettberg and his team at the University of Oregon are examining the differences between the seeds and DNA of wild chickpea plants collected from Turkey and Kurdistan and domesticated chickpea plants to better understand the genetic bottleneck created in domesticated chickpea plants and to exploit the characteristics of wild plants that allow them to grow in harsher environments. This effort has exploded, and these teams have joined together to examine over 400 wild chickpeas species in forty-five different countries, scouring their genomes for the genes that produce the traits that will optimize chickpea cultivation. So, chickpea enthusiasts, you might consider sending email to Dr. Agrawal, Dr. Varshney, and Dr. Bishop von Wettberg to thank them for their dedication and diligence! Few things inspire scientists more than recognition of their work.

DELECTABLE CHICKPEA DISHES

Roasted chickpeas, hummus, falafel, dried cooked chickpeas, bean salad with chickpea salad

Standard Recipe: Roasted Garlic and Red Pepper Hummus (Preparation time, 20–25 minutes; Cooking Time, 50–60 minutes)

Ingredients:

16-oz can of chickpeas

¼ cup of water

2 tsp lemon juice

2 tbsp tahini

2 tsp extra virgin olive oil

1 tbsp extra virgin olive oil

1 small garlic bulb

1 small red bell pepper

1 tsp salt

Directions:

1. Preheat the oven to 350° Fahrenheit.
2. Cut off the top of the garlic bulb and wrap the bulb in aluminum foil.
3. Pour 1 tsp of olive oil on the garlic bulb and sprinkle some salt on the garlic bulb.
4. Cut the red bell pepper in half and deseed, then place on parchment paper.
5. Place the parch paper on a baking sheet.
6. Pour 1 tsp of olive oil on the red bell pepper and sprinkle some salt.

7. Place the aluminum-wrapped garlic bulb on the same baking sheet and place inside oven to
8. bake at 350° for 45–60 minutes. Remove from oven and allow to cool.
9. Remove the garlic cloves from the bulb.
10. Roughly dice the roasted bell pepper.
11. Drain the can of chickpeas into a strainer.
12. Put the chickpeas, water, lemon juice, olive oil (1 tbsp), salt, roasted garlic cloves, and diced red bell peppers into a food processor.
13. Blend well and add more lemon juice, salt, garlic, or peppers to taste.

Nonstandard Recipe: Chickpea brownies (Preparation time, 15–20 minutes; Cooking Time, 25–30 minutes)

Ingredients:

1 15-oz can of chickpeas, drained and rinsed

½ cup of almond butter

½ cup of coconut sugar

1 tbsp melted coconut or avocado oil

1 tsp of vanilla extract

¼ cup of coconut flour

½ cup of cocoa powder

1 tsp baking soda

¼ tsp salt

2/3 cup of chocolate chips

Directions:

1. Preheat the oven to 350° Fahrenheit.
2. Blend the chickpeas in a food processor until smooth.

3. Combine the chickpeas, almond butter, coconut sugar, oil, and vanilla extract in a mixer until well combined.

4. In a separate bowl, combine the coconut flour, cocoa powder, baking soda, and salt, then slowly add to the mixer until the ingredients are well mixed.

5. Add the chocolate chips and mix for another 15 seconds.

6. Pour the mixture into a greased pan and bake at 350° for 20–25 minutes or until you stick a toothpick in and it's no longer wet when you pull it out.

What is a life without chocolate? This magic confection makes everything taste better and makes everyone happy. Milk chocolate contains milk and a lot of sugar and is less healthy. However, dark chocolate contains less sugar and is chock full of fiber, iron, copper, magnesium, manganese, potassium, phosphorus, selenium (antioxidant and pro-immune functions), zinc, polyphenols, and probiotics, those good (commensal) bacteria in our guts that help to optimize our digestive efficiency, immune defenses, and brain functions (and the interactions amongst the three). Even dark chocolate contains quite a bit of sugar, so moderation is key when consuming dark chocolate for health benefits. This is not what is happening.

CHAPTER 9

○ ○○○○○○ ○○○○○○○ ○ ○○○○○ ○○○○○○○

No! Not Chocolate!!

Humans consumed almost 80,000 tons of chocolate in 2016. This amount is expected to triple by 2030. This estimation is based on current availability of chocolate. However, a decline in availability and soaring prices will curb consumption. These factors are not included in the chocolate consumption equation, as far as I can discern.

How would anyone conceive of crushing and eating this stone of a bean?

Wait, what is chocolate? The word *chocolate* refers to the edible substance that is extracted from the cacao bean and can be traced back to Aztec word *xocoatl*, which refers to a bitter cacao drink with medicinal properties. The Latin name for the cacao tree is *Theobroma cacao*, which means "food of the gods." Prior to consumption, cacao beans were used for currency. The vast history of chocolate as a food is in a beverage form. Powdered chocolate, referred to as "cocoa" (don't be confused with the plant, cacao) originated in the Netherlands in 1828 and is produced by roasting and pulverizing cacao beans.

Francois Louis Cailler established one of the earliest solid-chocolate factories in Switzerland in 1819 after first tasting Italian chocolate at a local fair and then spending four years in Turin perfecting the craft of solid-chocolate production. Cailler, the brand, teamed up with Daniel Peter in 1875 to combine condensed milk with chocolate, and the first milk-chocolate bar was born, followed by rapid commercialization during the following decade. Hazelnut chocolate was added to the mix when Cailler and Peter teamed up with Charles Amedee-Kohler to form the Peter, Cailler, and Kohler Chocolats Suisses chocolate company, which was later purchased by Nestle in 1929 (export sales were acquired in 1904, enabling Nestle to sell this chocolate for the first time). Henri Nestle patented the process of making condensed milk powder and then replaced liquid condensed milk with powdered milk, greatly simplifying milk-chocolate production.

Rain makes chocolate, and chocolate can makeover a rainy day

It is estimated that a single chocolate bar requires 450 gallons (about ten bathtubs) of water to make. Very little of this water is needed in the processing phase. Most of this water is required by the live cacao plant. Cacao plants can only grow in intense heat with sufficient water to survive such heat. This heat/water balance is difficult to achieve, and these microclimates, all within twenty degrees' latitude north and south of the equator (a range of roughly

1,400 miles), will disappear in about fifty years (NOAA). Currently, over half of the chocolate produced for global consumption comes from just two nations, Cote d'Ivoire and Ghana. The IPCC reported Western Africa to be a hotspot for climate change. In the coming decades, cacao forests in this region will experience a dramatic reduction in rainfall and an average increase in ambient temperature of almost four degrees Celsius.

This may not seem like much, but it is enough to drive residual water out of the soil and away from the roots of cacao plants, essentially killing them. The IPCC predicts only 10 percent of existing cacao groves will still exist in 2050. Moving to higher altitudes can ward off some of the impacts of the temperature increase, but ground suitable for cultivation in the mountains of the Cote d'Ivoire and Ghana is rare at best because increased altitude equals less soil and more rock. Escaping to higher terrains has its limits.

Humans are not the only connoisseurs of chocolate

On top of this heat/water balance problem, there is relatively limited genetic variation in cacao plants. As a result, these plants can't mount sufficient immune responses to numerous pathogens, making them more sensitive to infectious disease. Two fungi are wreaking havoc on cacao plants. Witch-broom disease (caused by *Crinipellis perniciosa*) destroyed 80 percent of Brazil's cocoa plants in the early 1990s. Also, once limited to northwestern regions of South America, frosty pod rot (moniliasis disease, caused by *Monoliophthora roreri*) is present through much of South and Central America. Frosty pod rot destroys cacao beans from the inside out and typically kills about 80 percent of the infected crop. Both witch-broom disease and frosty pod rot are now poised to spread into Africa. These and other fungi (mentioned in Chapters 3, 5, 10, 13, 14, 15) are better able to expand their habitats on a more volatile planet because they require moisture to grow and reproduce, but then use drought and wind to spread their spores to new territories. Climate change is producing the exact types of weather patterns that facilitate fungal habitat growth and expansion.

So, how do we save chocolate?

Scientists have recognized that losing chocolate will have negative impacts on human health and quality of life and are using state-of-the-art genetic methods (clustered regularly interspaced short palindromic sequences, CRISPR) to modify the genes of cacao plants to permit them to thrive in their native habitats as they become hotter and drier (reported by Erin Brodwin, Business Insider, UK, 2017). One of the developers of the CRISPR method, Dr. Jennifer Doudna, along with Emmanuelle Charpentier at the Max Planck Institute in Germany, is actively engaged in plant modification in her laboratory at the University of California at Berkeley, called the Innovative Genomics Institute. Through her private company, Caribou Biosciences, Dr. Doudna has also licensed her technology to the chemical supergiant DuPont (part of the DowDuPont super-duper-giant) that supports DuPont Pioneer, directly aimed at improving farmer productivity globally.

Other private corporations are doing their part too. The Mars Corporation, a thirty-five-billion-dollar business, has teamed up with UC Berkeley, specifically with Myeong-Je Cho at the Innovative Genomics Institute, to investigate cacao plant disease management and the possibility of creating a disease-resistant cacao tree. Mars is also going green by separately investing one billion dollars to "Sustainability in a Generation," an internal effort to reduce their carbon emissions by at least 60 percent by 2050. Let's keep our hopes up for all of these efforts to save the cacao plant. Since this is perhaps the most important question in this book, I'll ask it again: "What is a life without chocolate?!?"

DELECTABLE CHOCOLATE DISHES

Chocolate, chocolate milk, chocolate milkshake, hot chocolate, chocolate brownies, chocolate cake, chocolate fudge, chocolate fondue, chocolate chip cookies, chocolate chip muffins, chocolate chip pancakes, chocolate-dipped strawberries (or any fruit), chocolate-covered hazelnuts (or any nuts), chocolate-covered crickets/ants/worms

Standard Recipe: Chocolate chip cookies (Preparation time, 15–20 minutes; Cooking Time, 12–15 minutes)
    Ingredients:
    1 cup of unsalted butter, softened
    1 ½ cups of brown sugar or coconut sugar
    2 large eggs
    1 tsp vanilla extract
    2 ¼ cups of all-purpose flour
    1 tsp salt
    1 tsp baking soda
    1 ½–2 cups of chocolate chips

Directions:
1. Preheat the oven to 375° Fahrenheit.
2. Combine the butter and sugar in a mixer until smooth.
3. Add the eggs one by one and mix well, then add the vanilla extract.
4. In another bowl, combine the flour, salt, and baking soda.

5. Slowly add the dry ingredients to the mixer and mix well.
6. Add the chocolate chips.
7. Scoop the dough into small balls and place them onto a cooking sheet with parchment paper.
8. Bake the cookies at 375° for 12–15 minutes.
9. Transfer to a wire rack so the cookies can cool completely.

Nonstandard Recipe: Overnight Chocolate Oats (Preparation time, 5–10 minutes; Chill Time, overnight)

Ingredients:
½ cup of rolled oats.
1 cup of plant-based milk
1 tbsp of chia seeds
1 tbsp of cocoa powder
3 tbsp of maple syrup

Directions:
1. Combine all ingredients into a mason jar and mix well.
2. Place jar into refrigerator and leave overnight.
3. Serve the following morning with chocolate chips sprinkled on top along with a splash of more plant-based milk if desired

Recently, in a correlational study performed by investigators at the National Cancer Institute and published in the Journal of the American Medical Association, baseline health measures were taken, and the number of cups of coffee people drink on a daily basis was counted. Physical health was re-assessed ten years later. I would have guessed more heart attacks associated with three cups of coffee per day. I was wrong. This study showed that people who drink two to three cups of coffee every day actually live longer!

CHAPTER 10

o ooooo ooooooo o ooooo ooooooo

How Will I Start My Day Without a Mug of Coffee?

Coffee contains pantothenic acid (vitamin B5, a food-to-energy conversion assistant), riboflavin, niacin, thiamine, magnesium, manganese, and potassium. Perhaps most importantly, coffee contains caffeine, which was designated as a drug by the Food and Drug Administration in the 1980s. When it reaches the brain, caffeine reduces the impact of the inhibitory neuromodulator adenosine, and so increases the levels of neurotransmitters that make us more alert, attentive, vigilant, and goal-driven.

Caffeine also causes vasoconstriction (a reduction in the diameter of the blood vessels) in the brain. Purified caffeine is a key component in most weight-control supplements because it reduces appetite and increases metabolic rate. Interestingly, caffeine increases fat-burning rate three times more in lean individuals as compared to obese individuals. So, the metabolic effects of caffeine are negatively related to fat content. I would have guessed the opposite: more fat, more fat burning. Anyway, it takes a half to a full cup of coffee to get my neurotransmitters moving in the morning, and if I don't drink coffee in the morning, I typically get a headache in the mid to late afternoon. Am I addicted? Likely so. How about you?

How would anyone conceive of roasting and brewing this stone of a bean?

The magical coffee bean (genus *Coffea*) can be traced back to ancient forests on the Ethiopian plateau and Sudan. Coffee cultivation began on the Yemeni district of the Arabian Peninsula in Sulfi shrines during the fifteenth century. Interestingly, the coffee beverage was roasted and brewed then in much the same way we prepare it today. In only 100 years, and without cars or airplanes or Facebook or Twitter, it spread throughout Persia, Egypt, Syria and Turkey. Coffee's reputation quickly spread even further, as pilgrims visited Mecca annually and took their stories back home, describing this new elixir as the "wine of Araby." In another hundred years or so, coffee houses became common throughout Europe, and coffee replaced beer and wine as the breakfast beverage. The word *coffee* was introduced into the English language via the Dutch *koffie*, which was preceded by the Ottoman Turkish *kahve* and the Arabic *qahwah*.

Coffee is currently grown worldwide. Over seventy countries produce coffee. But a bulk of the coffee consumed by humans is grown in Brazil, Vietnam, Sumatra, Columbia, Indonesia, and Honduras. The two most popular cultivars are *C. arabica* and *C.*

robusta. Americans can thank Starbucks for making *C. arabica* so popular as that is their bean of choice and now nearly every commercial blend is made with the *C. arabica* bean. *C. robusta* has a higher caffeine content and is less expensive than *C. arabica* and has, possibly unjustly, earned the reputation as a second-rate coffee.

Cacao plants are doomed by drought and extreme weather events

As mentioned above, coffee is grown all over the place. So, we might think that coffee would be relatively protected against the effects of climate change. However, coffee is a volatile crop, and about 60 percent of wild coffee plants are at risk of extinction. One-third of the total coffee and half of the *C. arabica* bean consumed today is grown in Brazil, which will become warmer and drier in the future, two variables that downgrade coffee plant health and yield. The same holds true for the second and third-largest producers of coffee, Central America and East Africa. The global area suitable for coffee plant growth will be cut in half by 2050, according to studies by the International Center for Tropical Agriculture (CIAT).

Normally, *C. arabica* grows at higher altitudes and produces a reduced yield compared to *C. robusta*. *C. arabica* will be more susceptible to the consistent increases in local temperatures. *C. robusta* will be more sensitive to large fluctuations in temperature and precipitation. As recently as 2014, crop yields were slashed by roughly 20 percent due to droughts occurring in Brazil and Vietnam. In the same year, more extreme weather in Central America facilitated the growth of a leaf rust fungus (first appearing in east Africa about 150 years ago) that initially robs the leaves of sunlight and then chews holes in them as it makes its way to the beans to destroy them. As severe weather fluctuations continue or worsen, 40 percent of all coffee plants grown in Central America are expected to disappear due to rust fungus alone. Moreover, both coffee types will be more susceptible to tiny black beetles known as "coffee berry borers" (*Hypothenemus hampie*), which are spreading as the global temperature increases. None of this is good news for coffee plants or coffee drinkers.

Losing coffee may have a greater impact on human civilization than all other foods described in this book, even chickpeas and consideration of the hundreds of millions of individuals who rely on chickpeas for protein and health. Coffee is addictive and enables a majority of adults on this planet to start the day and to deal with daily stress. Reduced coffee availability combined with caffeine withdrawal and climate change–driven increases in physical, social, and political stress is a recipe for disaster. There is some hope for the future of coffee given human activities directed at saving the plant. The World Coffee Research Program in Portland, Oregon, and the Texas A&M Coffee Center are investing heavily to support coffee producers in Central America, Mexico, Puerto Rico, South America, and Africa. They also conduct genetic studies to enhance coffee plant health, generate rust-resistant coffee plants, and create a sustainable coffee industry in the face of climate change. Once again, geneticists are called upon to save the day.

DELECTABLE COFFEE DISHES

Cappuccino, espresso, latte, mocha, iced coffee, coffee cake, mocha muffins, coffee sugar cookies, tiramisu, affogato, coffee-rubbed meats

Standard Recipe: Affogato (Preparation time, 5 minutes; Brewing Time, 3–5 minutes)
Ingredients
3 scoops of your favorite vanilla ice cream
1/3 cup of strong brewed coffee

Directions:
1. Place the 3 scoops of vanilla ice cream into a bowl or cup.
2. Pour the coffee on top.

Nonstandard Recipe: Coffee Dry Rub Ribs (Preparation time, 4 hours 20 minutes; Cooking Time, 2 hours 5 minutes)
Ingredients:
1 tsp ground coffee
1 tbsp salt
1 tbsp ground cumin
1 tbsp garlic powder
1 tbsp onion powder
1 tbsp brown sugar
1 tsp oregano
1 tsp ground black pepper
2 racks of baby back ribs

Directions:

1. Combine all ingredients except for the ribs into a small bowl and stir.
2. Rub the dry ingredients onto the ribs and cover tightly with aluminum foil.
3. Marinate for at least four hours.
4. Preheat the oven to 250° Fahrenheit.
5. Place the covered ribs onto a baking sheet and into the oven.
6. Cook at 250° for 2 hours.
7. Switch the oven to broil and uncover the ribs.
8. Broil until the ribs darken.
9. Take the ribs out of the oven and let them rest for 15 minutes.
10. Serve as is or with your favorite barbeque sauce.

I think my personal history with fish as a food is pretty typical for an American kid. I didn't care much for fish as a young kid, unless it was breaded, deep-fried, accompanied by French fries, and served with a bucket of ketchup. A liking for cooked fish grew in me as a young adult. Sushi was a taste acquired a little later in life. Now, the simple salmon roll is my favorite type of fish and in the top five of my favorite of all types of foods. Just a piece of raw salmon and seasoned rice wrapped in a paper-thin sheet of nori (shredded seaweed pressed back together) and then dipped in soy sauce infused with wasabi. Yum!

CHAPTER 11

○○○○○○○○○○○○○○○○○○○○○○○○○○

Fish Are Delicious Food Chain Linchpins

The situation was completely different for my wife, who was raised in Vietnam for the first four years of her life before immigrating to the Unites States. Fish became a major component of her diet as soon as she started eating solid foods. She thinks broiled mackerel was her first fish. Her mom says it was congee (a rice porridge) with Cotton fish when she was just a year old. Her uncle remembers preparing Snakehead casserole (*Cá Lóc*) with fish sauce (*Cá Kho*) for her when she was five years old.

Southeast Asians put fish sauce on everything. Fish sauce is made from anchovies, salt, and water that is fermented and then slowly pressed to release all of the liquid. From this person's Caucasian perspective, fish sauce smells horrible, but tastes wonderful. How can that be when a large portion of our taste perception is guided by smell? If you want the answer, write to Dr. Peter Brunjes, a neuroscientist at the University of Virginia. He includes an awesome jellybean demonstration in one of his undergraduate sensory perception courses to illustrate the interactions between smell and taste.

Fish are delish and mostly healthy to eat

Many of the fish we eat contain numerous minerals (calcium, iron, iodine, magnesium, phosphorus, potassium, and zinc) and vitamins (B2, B6, D) that promote better health. Perhaps the most popular understanding of the health benefits of eating fish is that many types of fish contain omega-3 fatty acids, which act as antioxidant and anti-inflammatory molecules in our bodies. Higher levels of omega-3 fatty acids are associated with improved mental health, coronary health, vision, and pregnancy outcomes. However, omega-3 does not act alone in regulating inflammation, and omega-6 fatty acids also need to be considered. It is thought that the omega-6 to omega-3 ratio in the bloodstreams of primordial humans was somewhere between one to one (1:1) and 5:1.

Currently, omega-6 to omega-3 ratios in Western civilizations range from about 10:1 to 30:1! A higher omega-6 to omega-3 ratio is associated with a "Western diet" consisting of red meat, refined grains, sugar, and salt. Mammals lack the required enzyme to convert omega-6 to omega-3. So, the shift from grass to grain as cattle feed to fatten the animals more rapidly has increased the omega-6 to omega-3 ratios in these animals, contributing to the higher omega-6 to omega-3 ratios in our diets and our bodies. An increased omega-6 to omega-3 ratio has been associated with increased systemic inflammation, which can increase predisposition to a variety of "Western diseases," including coronary disease, stroke, diabetes, and breast cancer. However, there is also evidence that increased intake of arachidonic acid does not increase measures of inflammation in healthy human subjects. Clearly, more research on omega-3, omega-6, and the not-so-distant cousin omega-9 fatty acids are needed to clarify their interactions with each other and roles in human health and disease.

Fish can't withstand both human and climate attacks

Unsustainable practices have plagued the fishing industry for decades. Some of these bad practices have to do with the method, and some are simply a function of overfishing. Poor management of fisheries, such as lack of enforcement of regulations and too few no-go fishing zones, has resulted in unwarranted depletion of numerous fish types, including the Atlantic halibut, the bluefin tuna, the monkfish, and various whales (numbers at GreenPeace. org). Secondary impacts can be seen far north as well where overfishing by Alaskan pollack fisheries are causing a reduction in the number of already endangered northern fur seals who rely on pollack as their major food source.

Pirate fishers that do not respect established fishing restrictions and also over-catch juveniles are reducing reproductive capacity and population sizes of future generations. Moreover, unfair business deals made with developing nations by "legitimate" fishing companies has enabled exploitation and overfishing in their coastal territories. On top of this mess, innocent bystanders are very often killed by bad fishing practices and are then simply discarded, including birds, dolphins, sea turtles, sharks, and corals. As such, the fishing industry is already struggling, and the climate problem has yet to wreak its ocean deconstruction in full force.

The oceans have been warming rapidly due to human energy-consumption activities. This is not directly evident to most off-the-rack human beings who can't see temperature with their eyes, typically feel the ocean as cool or cold to the touch, and don't live there (not at the beach in a five bedroom cottage—in the ocean, under the water). Way back in the 1960s, it was reported that almost 30 percent of marine species were already disappearing. More recent research poses grim predictions for the future. Here's the deal: If we don't tackle this problem now, there will be NO SEAFOOD AT ALL in fifty years. This is not new information. An important article in the field was published back in 2006 in *Science Magazine* by a research team out of Dalhousie University that warned of the harm to ocean health caused by climate change and a loss of biodiversity. Their research predicted very dire consequences for our ocean ecosystems that are being borne out in real time.

Overall, without additional and dramatic action, ocean health will continue to deteriorate at an accelerating rate. If there are minimal living creatures in our seas and oceans, then all of those animals that rely on marine life for food will also perish, like seals, whales, dolphins, seagulls, polar bears, and many other species. There are huge ramifications of this for human beings, and there is no rapid return. It took thousands of millions of years to create the oceans' ecosystems. They cannot be rebuilt. They have to be slowly re-evolved. Without bold actions to improve ocean health now, anyone reading this will be long dead before the oceans even begin to recover.

Fish that can't adapt are more likely to perish

Marine species referred to as "generalists" may be least affected since they are not particular in the food types they eat and can adapt to various habitats. "Specialists," those animals that have limited mobility and food sources, will suffer the most. From a NOAA study of eighty-two saltwater animals on the Northeast

US continental shelf, it has been estimated that about 17 percent are temporarily benefitting. Examples include anchovies, black sea bass, bluefish, and Spanish mackerel. Twenty-eight percent are at minimal risk, including the dusky shark, porbeagle, sand tiger, and spiny dogfish. Forty-three percent are already showing negative impacts, such as the Atlantic cod, Atlantic halibut, Atlantic salmon, Atlantic sea scallop, blue mussel, eastern oyster, and winter flounder.

Scallops and salmon are amongst the most susceptible fish to increasing global temperatures. To go off track for a moment, as part of metabolism in our bodies, CO_2 is produced, and we exhale it so it doesn't accumulate. If it does, blood acidity increases due to the formation of carbonic acid. Returning to the topic at hand, the same is true of CO_2 in the air and acidity in our oceans. More CO_2 in the air equals more carbonic acid in the ocean. Increased ocean acidification hinders shell development and movement in scallops, making them more vulnerable to predators. Other shellfish, like clams and oysters, will also be impacted by ocean acidification. If unchecked, it is fairly safe to assume that climate change will cause most types of shellfish and corals to perish before the turn of the century. A 50 percent loss in scallop yield is expected by 2050.

Salmon spend part of their lives in smaller bodies of water that warm up more than the open ocean, and so salmon are more temperature sensitive than many other fish species. As rivers become warmer, the water can hold less oxygen, which the salmon need to breathe. The warmer water also makes them more susceptible to parasites and disease. At least three species of salmon of the Great Northwest are now considered endangered as a direct result of increasing water temperature of rivers and estuaries in Washington State: the spring Chinook salmon, the summer steelhead trout, and the bull trout. Salmon and trout belong to the same fish family, *Salmonidae*, but spend different amounts of time in fresh versus salt water. Ocean acidification is negatively impacting the food supply for salmon, which includes krill and pteropods, also known as sea butterflies. Also, as sea levels rise, more salt water is pushed inland, altering the river and estuary hydrology balance and further exacerbating the salmon survival problem (similar to the effects of Potomac River overflow into the Washington DC tidal basin increasing salinity at the roots of cherry trees).

The sockeye salmon of Alaska's Bristol Bay may have it the worst. Unlike other salmon species, these poor fish have a "blob" of warm seawater in the North Pacific Ocean on one side and land on the other. The blob is a shifting mass of warm water, up to seven degrees warmer than normal. This blob was first detected in 2013 and given the name by Nick Bond, a research meteorologist at the Joint Institute for the Study of Atmosphere and Ocean in Seattle, Washington, which is supported by the University of Washington and NOAA. The sockeye live in a pocket of coastal water that is cooled by coastal upwells from the ocean depths. Both the blob and the beach are warmer than this cool zone. Since salmon use temperature as an important navigation cue, when ready to spawn, the sockeye get confused and don't know which way to go to head upstream. On top of this, creation of a proposed hydroelectric dam at Chikuminuk Lake in Wood-Tikchik State Park and what would be the largest open-pit mine in North America at the Bristol Bay headwaters further threaten the delicate ecosystem that allows the sockeye to complete its lifecycle. These are only a few examples of salmon distress. The complete list of problems for the salmon is long, and these poor fish are in a serious quandary.

This fish problem is too huge for simple fixes

One might think that efforts to save salmon would include not building dams and mines that destroy the habitats in which they live and breed, but they do not. Important efforts are being made to try to preserve the river systems required by salmon during the breeding season since these rivers are where they are most endangered and river waters can be addressed more easily than ocean waters. However, this problem is so massive that specific plans to save salmon are simply insufficient. The conversation quickly shifts to the fulfillment of regional and global commitments to curb climate change, including reducing carbon emissions in the Pacific Northwest and investing in carbon-neutral and sustainable energy resources, primarily wind and solar energy technologies.

DELECTABLE FISH DISHES

Fish and chips, seared tuna, sushi, blackened catfish, broiled snakehead fish, grilled salmon, congee with fish, steamed fish with garlic and ginger

Standard Recipe: Grilled salmon (Preparation time, 35–40 minutes; Cooking Time, 10 minutes)
Ingredients:
2 salmon fillets
1 tbsp garlic powder
Salt
Pepper
1 lemon
1 tbsp avocado oil

Directions:
1. Place the salmon fillets on a plate and marinate by sprinkling salt, pepper, and garlic powder.
2. Let the salmon marinate for at least 30 minutes.
3. Heat the avocado oil in a large pan.
4. Carefully place the salmon fillet into the pan.
5. Cook the salmon for 10 minutes while flipping midway.
6. Place the salmon on a plate and squeeze the lemon on top.
7. Serve with steamed rice and vegetables.

Nonstandard Recipe: Vietnamese Fish Sauce (Nuoc Mam) (Preparation time, 1 minute; Cooking Time, 5 minutes)

Ingredients:

1 cup of fish sauce

1 cup of sugar

4 cups of water

Sliced chili peppers (optional)

Directions:

1. Pour the water, sugar, and fish sauce in a sauce pan and turn on the heat to medium low.
2. Stir until the sugar dissolves in the liquid.
3. Throw in the sliced chili peppers.
4. Serve hot or cold as a dipping sauce.

There is no fiber or fat in honey. It is mostly fructose and glucose, with lesser amounts of maltose and sucrose. It is produced primarily from flower nectar that honeybees (genus Apis) drink and then regurgitate. That's correct. Honey is honeybee vomit. The act of ingesting mixes the nectar with enzymes in the bees' guts that alter the sugars. One enzyme, invertase, breaks down complex sugars (disaccharides) like sucrose into simple sugars (monosaccharides) like glucose and fructose. Glucose and fructose taste much sweeter than sucrose. Regurgitation also permits evaporation of water and thickening of the product.

CHAPTER 12

○ ○○○○○○ ○○○○○○○ ○ ○○○○○ ○○○○○○○

Winnie-the-Pooh Is Having Panic Attacks About Honey, "Oh, Bother"

Meli is the Greek root word for honey (but can be generalized to anything with a sweet taste). Melons get their names from this derivative. Loosely translated, "Melissa" means honeybee. My sister's golden retriever earned the name "Meli" due to her honey-colored fur coat and sweet disposition.

Honey is more than just sugar

A vast majority of the honey we eat comes from beekeepers (apiculturalists) that maintain domesticated bees in man-made hives. A lesser amount of honey is harvested from wild honeybee hives. Health wise, raw unprocessed honey contains organic acids and phenolic compounds that act as antioxidants. The variety of these molecules found in honey is thought to produce a heightened antioxidant capacity. Some of these organic acids with proven antioxidant properties are acetic (vinegar), butyric, citric (vitamin C), formic, gluconic, lactic, malic, pyroglutamic, and succinic acid. The antioxidant phenolic compounds in honey are primarily flavonoids, which are plant pigments typically orange, red, or blue in color. Honey also has some antibiotic properties and has been show to kill the bacteria that cause staph infections (these are chest colds and skin infections caused by the bacterium *Staphylococcus aureus*).

The high levels of sugar would be the major concern when eating a lot of honey, but ounce for ounce, honey is more healthy to eat than refined sugar because it takes a bit more time to get through your stomach to your small intestine, where it is absorbed, and so has a lower glycemic index (ranging from about forty-five to sixty-five). Note that honey may not be an appropriate sugar substitute for diabetics due to its methylglyoxal content. Methylglyoxal alters low-density lipids (think "bad" cholesterol) so that they accumulate more easily on cell surfaces, potentially worsening cardiovascular health and exacerbating neuropathy in people suffering diabetes.

Who was the lucky guy first assigned to harvest honey from wild honeybee hives?

The exact origin of human consumption of honey remains uncertain. Some experts say the honeybee itself originated in southern Asia, while others say Africa. A 100-million-year-old honeybee was discovered embalmed in amber in what is now Myanmar. Eight-thousand-year-old cave paintings in the Cuevas de la Araña ("spider caves") in the municipality of Bicorp in Valencia, Spain, depict the climbing of trees and harvesting of honey from wild honeybee hives. Honey is also mentioned in the Vedas (religious texts) and Ayurvedic texts (alternative medicine practices) of ancient India that are at least 4,000 years old. Based on actual honey remains, the earliest human consumption can be dated back about 5,000 years to Georgia, on the Europe-Asia border, where honey remains have been detected inside clay pottery excavated from ancient tombs.

Honey was also used as a topical medicine to treat skin injuries and rashes and, when fully hardened, as a talisman (a gem or charm possessing mystical power). The twelve gods of Olympus in ancient Greek culture were said to eat honey in the form of ambrosia. There were so many domesticated honeybee hives in ancient Greece (circa 2,600 years ago) that a social-distancing law was passed such that honeybee hives had to be separated by roughly 100 meters (about 300 feet). In Hinduism, priests pour honey (called *madhu*) over symbols of the deities as an offering while they chant mantras during "madhu abhisheka," a devotional celebration to the Hindu deities. Honey also is considered one of the five elixirs of Hindu life. The Hebrew bible also makes numerous references to honey as an offering to the Lord and as a symbol of the Promised Land. On a more contemporary level, there is perhaps no vinyl album cover more sexually provocative than that of "Honey" by the Ohio Players, released in 1975 (google it).

Bee endangerment is not a new problem

Okay, this chapter is about honey (yum!), but it is also about bees because bees produce honey. Before diving directly into the effect of honeybee loss on honey, a bigger issue is the loss of all types of pollinating bees on plant-based food availability in general. Plants don't hug and kiss and have foreplay before they engage in physically intimate intercourse. Something has to move pollen (semen) from the male plant's stamen to the female plant's carpel, which contains the ovary. Bees do this better than any other animal on this planet. (To a lesser extent, some other bugs, birds, and mammals act as pollinators.) Bees are the primary pollinators of 80 percent of the world's plants. Conservatively, bees pollinate the crops that produce about 35 percent of the commercial vegetables, grains, and fruits that we eat. We can't lose bees!

Bees have been dying off at record rates for about a decade and are threatened most by invasive parasites, reduced food quality due to reduced nutrient content in domesticated crops and mono-cropping farming practices, and man-made chemicals (mostly neonicotinoids), which kill them outright or rob them of their navigation abilities. I haven't even mentioned any impact of climate change yet. The impact of climate change on bees precedes these more recent issues (pestilence, nutrient limitation, poisoning). The southern borders of bee habitats in the United States and in Europe are receding at about five miles per year, and have shifted north by about 200 miles since the 1970s. As important, the northern border of bee territories has not shifted further north, likely due to a variety of factors, including reduced daylight length, food availability, and restricted expansion ranges for small colonies. The result of this territorial squeezing is a continual shrinking of bee habitat sizes for the better half of a century.

In addition to a reduction in habitat area, climate change causes many wild and domesticated flowering plants to bloom early, causing ecological scientists to worry about the timing between pollinator presence and pollination readiness in flowering plants. Rebecca Irwin and her team at North Carolina State University have looked at the impacts of time of flower blooming on pollinator behavior and plant health. Through longitudinal studies lasting many decades, they showed that bee health is directly tied to flowering plant health. They also found that flowering occurring earlier in the spring due to climate change may actually benefit the plants by increasing interaction time with pollinating bees. However, the same research project showed that this comes at a cost because early pollination makes the plants much more susceptible to subsequent frosts, which are typically deadly. This is troubling news for plants and pollinating bees alike.

Numerous nations are taking actions to identify the most pronounced causes endangering bees and developing ways to mitigate these problems. The World Wildlife Federation and Buglife have teamed up to fund and perform studies on bee population statistics, bee health, bee habitats, and the impact of climate change and other factors that can be remedied by alterations in human behavior. With immediate and appropriate action, there is hope for bees in general, for the honeybee in particular, and for the future of honey and all plants requiring pollination.

Note: Before setting up a honeybee hive in your backyard, talk to your local environmental expert about the impacts on native bee populations. If you google "suburban wildlife services" or "suburban wildlife management," you will get hit after hit on how to kill things. Don't do this. I also typed in "Fairfax bee expert" (I live in Fairfax, VA) and still got a long list of exterminators and beehive relocators. You need to include "bee" and "garden" and "expert" to begin to get meaningful information and helpful sources.

DELECTABLE HONEY DISHES

Honey corn muffins, honey granola, honey granola bars, honey-glazed pretzels/carrots/ham, baklava, honey taffy, tea with honey, apples dipped in honey, honey mustard dipping sauce

Standard Recipe: Honey granola (Preparation time, 10 minutes; Cooking Time, 25–30 minutes)
    Ingredients:
    ¼ cup of avocado oil
    ¼ cup maple syrup
    ¼ cup honey
    1 tsp ground cinnamon
    1 tsp vanilla extract
    3 cups of old fashioned rolled oats
    ¼ cup chopped walnuts
    ¼ tsp salt

Directions:
1. Preheat the oven to 325° Fahrenheit.
2. Line a baking tray with parchment paper.
3. Combine all ingredients into a bowl and stir until the oats are covered with the spices, syrup, and honey.
4. Spread the mixture evenly on the baking tray and bake at 325° for 25–30 minutes, stirring occasionally until the oats are golden brown.
5. Remove from oven and let it cool for 10 minutes.
6. Store in an air-tight container.

Nonstandard Recipe: Baklava (Preparation time, 30 minutes; Cooking Time, 1 hour and 20–30 minutes)

Ingredients: Syrup

2 cups sugar

1 ½ cups water

1 cinnamon stick

Juice of ½ a lemon

2 lbs filo dough

1 tbsp unsalted sweet butter

2 lbs crushed walnuts

2 tbsp sugar

¼ tsp nutmeg

¼ tsp cinnamon

16 oz honey

Directions:

1. Boil syrup ingredients until bubbles become large (10–15 minutes). Let cool.
2. Melt butter on low heat and use a butter brush to coat the bottom and sides of a 12 x 16-inch pan.
3. Preheat oven to 350° Fahrenheit.
4. Mix together walnuts, sugar, nutmeg, and cinnamon in a large bowl.
5. Place 6–8 sheets of filo dough flat on the bottom of the pan, and butter the top sheet.
6. Use large spoon to apply half of the walnut/sugar/spice mix, and spread evenly over filo.
7. Place 4 sheets of filo dough flat on the walnut/sugar/spice layer, and butter the top sheet.

8. Apply remaining walnut/sugar/spice mix, and spread evenly over filo.
9. Place 4 sheets of filo dough flat on the walnut/sugar/spice layer, and butter the top sheet.
10. Place 1 sheet of filo dough on top and do not butter.
11. Use a sharp bread knife to cut the uncooked baklava into diamonds.
12. Bake at 350° for 20 minutes, then lower the temperature to 275° and cook for another hour.
13. Remove from oven and apply honey to all of the cracks between the diamond pieces while hot.
14. Let cool for 10 minutes, then pour syrup into all of the cracks between the diamond pieces.
15. Let set overnight before removing pieces from the pan.

Rarely are peanuts eaten raw. Mostly, they are roasted and salted or ground into peanut butter. Peanuts are high in protein and fiber and low in carbohydrates, so they have a low glycemic index. Peanuts also contain choline, folate, niacin, pantothenic acid, riboflavin, thiamin, vitamin B6, and vitamin E, and are rich in minerals like magnesium, phosphorus, potassium, zinc, iron, copper, manganese, and selenium. They also contain far more unsaturated (good, vegetable) than saturated (bad, animal) fats.

CHAPTER 13

o ooooo oooooooo o ooooo oooooo

When Life Stops Giving Peanuts, We Stop Making Peanut Butter

Peanuts contain a lot of oil, and a large proportion of peanuts are grown to produce peanut oil for cooking purposes. Also on the healthy side, peanuts contain a lot of biotin, which is a beneficial molecule in a variety of ways. On the not-so-healthy side, peanuts also contain the proteins arachin and conarachin, which can cause severe and life-threatening allergic reactions in some individuals.

About 700 million tons of peanut butter is consumed annually. This equates to three pounds per person per year. Barring those with allergies, nearly everyone has eaten a peanut butter sandwich, either alone or with some sort of jelly or even marshmallow cream, known commercially as Fluff. You can make marshmallow cream with two fingers, two thumbs, and a marshmallow. Just keep pinching and pulling and folding and pinching and pulling and folding, and eventually it takes on a softer-than-taffy-like texture. But I digress. As a young adult, I may have eaten about three pounds of Reese's peanut butter cups per year, mostly around Halloween. Since I'm older now and no longer immortal, peanut butter or peanut butter and jelly sandwiches have largely replaced the candy, but I probably ate more than a pound of peanut butter in the last year. How about you?

Nobody really knows where the peanut originated

The earliest archaeological evidence of the peanut plant (*Arachis hypogaea*) is in Peru, where digs have uncovered peanut pods estimated to be about 7,500 years old. However, peanuts were likely domesticated earlier in wetter places that would be void of the same archaeological evidence found in drier climates like Peru. Evidence for this comes from discoveries that pre-Columbian cultures in South and Central America, such as the Mayans and Moche, depicted peanuts in their art. Peanut cultivation expanded from South America to Spain as explorers brought peanuts home with them. From Spain, peanut cultivation then spread in multiple directions, including further east to Asia and south to Africa. From Africa, peanuts traveled back across the Atlantic Ocean to America. This is kind of a roundabout way for peanuts to get to North America from South America. Peanuts are now grown in tropical and subtropical regions across the globe.

Peanuts were first introduced to Americans in large numbers at the 1904 World's Fair in St. Louis, Missouri, and reinforced as an important nutrition source by its use in the two world wars. Peanuts have earned a special place in American history thanks in large part to Dr. George Washington Carver. While born into slavery in Diamond, Missouri, it seems Dr. Carver was also born into botany as well and was labeled the "plant doctor" by local farmers as a young child. One of his major contributions to agriculture in the US was to increase the awareness of the dangers of soil erosion, and for this reason he pushed for supplementation and replacement of cotton with other staple crops. Peanuts rose to the top of the list as a rotation crop because the decomposition of peanut plants returned to the soil what cotton plants depleted, especially nitrogen. Cotton crop yields soared and farmers were elated. To deal with massive surpluses of peanuts, Dr. Carver eventually created roughly 300 products based on peanuts, not all edible, including flour, paste, paper, soap, shaving cream, and skin lotion. With the advent of machinery to facilitate harvesting and shelling, peanut cultivation rapidly expanded to the point where the United Peanut Association of American was formed in 1920.

Peanut consumption has increased considerably since the turn of the century, mostly in the form of peanut butter and less in the form of unshelled peanuts. Only three nations eat more peanuts than the US—China, India, and Nigeria. Only two nations produce more peanuts than the US, China and India. There are four basic types of peanuts grown in the US, the Runner, the Spanish, the Valencia, and the Virginia peanut. Runners are grown primarily in Alabama, Florida, Georgia, with a majority of these peanuts churned into peanut butter. Spanish peanuts are grown in Oklahoma and Texas. Valencia peanuts are grown in New Mexico, and Virginia peanuts are grown in Virginia and the Carolinas. And while we eat more peanuts than we ever have, the price of peanut products has also skyrocketed due to climate change. Sold at about two dollars per pound in 2011, peanut prices are now about 30 to 40 percent higher.

Peanuts are the Goldilocks of legumes

While most legumes have their beans above ground, the

pods of a peanut plant develop underground. Peanut plants are delicate and shy and need things just right. The margin for error between a healthy peanut plant and a dead peanut plant is small. They can't be planted too early because they are highly sensitive to late winter frost. Moreover, peanut plants require about five months of consistently warm weather and quite a bit of water (twenty to forty inches during the growing season) that tapers in amount as harvest approaches. If water availability drops too slowly, mold can develop and peanuts are hard to pick. If water availability drops too quickly, pods don't germinate and the plants become contaminated by an aflatoxin produced by soil fungus, which can be deadly if consumed in high quantities.

This environmental situation happened in 2011 when a major summer drought hit the southern and southeastern portions of the US, causing many peanut plants to shrivel up and die, producing no peanuts at all. This was not a temporary event; peanut production decreased from about seven billion pounds to about five and a half billion pounds in just one year, from 2017 to 2018. Conservatively, about 25 percent of peanut species will become extinct by 2030. By 2050, nearly all gone. No unsalted peanuts. No honey roasted peanuts. No caramel popcorn and peanuts. No peanut butter and jelly sandwiches. No Reese's peanut butter cups, including Reese's peanut butter chocolate Easter eggs that have the perfect balance of milk chocolate, peanut butter, and sugar to mesmerize and completely satisfy any peanut confection lover. My soul cried a little bit while writing that last sentence.

Peanuts teach us about carbon sequestration

Adding charcoal as a soil amendment has been shown to reduce greenhouse gas emission (methane, not CO_2) from peanut crops and increase peanut yield. In this soil amendment form, the charcoal is called "biochar." Biochar is created through pyrolysis, the heating of biomass in the absence of oxygen, which prevents combustion. Decomposition of biomass in the Amazon Basin produces a black soil called *terra preta*, which means "black soil" and is high in biochar content and very fertile. In general, biochar may be an important player in the movement of carbon out of the air and back into the ground where it can be stored for a very long time without impacting global temperature. Development of such carbon-negative technologies are being promoted by serious brainiacs. Dr. James Hansen, former director of NASA's Goddard Institute for Space Studies and current director of the Program on Climate Science, Awareness, and Solutions at Columbia University's Earth Institute, strongly endorses this approach, along with greater investment in nuclear energy.

Dr. James Lovelock is the inventor of the electron capture detector that enabled studies of the impact of chlorofluorocarbons (CFCs) on stratospheric ozone levels. He believes the answer to global warming is growing more food to capture CO_2 in the plants and then converting the parts of the plants we don't eat into charcoal that can be buried. Dr. Lovelock is also the creator of the Gaia hypothesis that proposes that living and nonliving parts of our planet should be viewed as one organism. Climate change illuminates this hypothesis. All living things are interconnected and dependent on the composition of the nonliving parts of Earth. When you write to your state and federal representatives, as suggested in the final chapter, you can sound like a brainiac yourself by mentioning these two people and promoting investment into technologies that permit sequestration of CO_2 and solidification of carbon.

DELECTABLE PEANUT DISHES

Roasted peanuts (salted or unsalted or caramelized), peanut butter, peanut butter and jelly (or fluff) sandwich, peanut brittle, Thai peanut chicken

Standard Recipe: Peanut brittle (Preparation time, 10 minutes; Cooking Time, 30–35 minutes)
  Ingredients:
  2 cups sugar
  1 cup corn syrup
  ¼ tsp salt
  ½ cup water
  1 cup butter
  2 ½ cups dry-roasted unsalted peanuts
  1 tsp baking soda

Directions:
1. Mix together the water, corn syrup, and sugar in a deep sauté pan.
2. Cook at medium heat stirring frequently until all of the sugar is dissolved and the mixture boils.
3. Add the butter to the water/syrup/sugar mixture and continue to cook while stirring frequently for 15 minutes.
4. Add the peanuts to the mixture and continue to cook at 300° while stirring frequently for another 10 minutes.
5. Turn off the burner and stir in the baking soda.
6. Place a piece of parchment paper on a cookie tray.

7. Pour peanut brittle mix onto the parchment paper and spread evenly to about a ¼ inch thickness.
8. Let cool for 25–30 minutes and break into small shards.

Nonstandard Recipe: Hotteok (Korean Pancake) with Peanuts and Cinnamon (Preparation time, 3 hours and 10 minutes; Cooking Time, 3–5 minutes per hotteok)

Ingredients:

1 cup of all-purpose flour
1 cup of sweet rice flour
¾ tsp salt
1/3 cup of water
1 packet of active dry yeast
¾ cup of water
1 tsp of sugar
½ cup of unsalted peanuts
½ cup of sugar
1 tbsp of ground cinnamon
10 tbsp of avocado oil

Directions:

1. Add the active dry yeast into ¾ cup of warm/hot water, add the 1 tsp of sugar and stir well. Let the yeast water sit for 10 minutes.
2. Combine the flour, sweet rice flour, and ¾ tsp of salt into a medium bowl. Pour the yeast water into the same bowl and mix well.
3. Let the wet dough sit for 3–4 hours as it will double in size.

4. Combine the unsalted peanuts, ½ cup of sugar, and cinnamon in a small bowl.
5. Pour the avocado oil into a frying pan over medium heat.
6. Pour some oil into the palm of your hands and rub.
7. Pour about 3 tbsp of the wet dough into one of your hands and spread it into a circle about 3–4 inches in diameter.
8. Put ½–1 tbsp of the sugar peanut mixture into the wet circle of dough and close it by gathering the corners.
9. Place the closed dough into the frying pan and cook it on one side for 1–2 minutes until the sides turn golden brown.
10. Flip it onto the other side, then flatten it using the spatula. Cook for 1–2 minutes until the sides turn golden brown.
11. Remove from the pan and let it cool for a few minutes before enjoying.
12. Repeat steps for the remaining wet dough.

The pomme de terre, or "apple of the earth." Its real name is Solanum tuberosum. Unlike many of the foods discussed in this book (with the exception of chickpeas), potatoes are a staple crop on which millions of individuals rely on for survival. Potatoes are more than starch energy. They contain a host of health nutrients and minerals, including vitamin B6, vitamin C, folate, magnesium, manganese, niacin, phosphorus, and potassium. Like honey, potatoes also contain antioxidants, including carotenoids, flavonoids, and phenolic acids.

CHAPTER 14

○ ○○○○○○ ○○○○○○○ ○ ○○○○○ ○○○○○○

The Potato Was the Ultimate Comfort Food

When potatoes are heated above 250° Fahrenheit, acrylamide is formed. For comparison, French-cut potatoes are typically fried at temperatures at or above 325°, and whole potatoes are baked at 400°. In a molecular biology laboratory, acrylamide is set into polyacrylamide gel that is used to separate proteins by size after extraction from a biological sample. In our bodies, acrylamide is converted into glycidamide, which is toxic and may act as a carcinogen.

Laboratory experiments have produced solid evidence that acrylamide increases cancer risk, and both acrylamide and glycidamide were deemed probable carcinogens by the International Agency for Research on Cancer in 1994. Acrylamide is also classified as "likely to be carcinogenic to humans" by the US Environmental Protection Agency and "reasonably anticipated to be a human carcinogen" by the US National Toxicology Program. Acrylamide can be avoided by not storing potatoes in the refrigerator and by soaking potato slices in cold water for at least a half an hour before cooking. Recipes for quality French fries and potato chips call for soaking the potato slices for a couple of hours before frying, so we're good to go, if we cook our French fries at home. Also, potatoes can be successfully fried or baked at temperatures below 250° Fahrenheit.

The potato is the great-grandpa of all tubers

Inca Indians in Peru and Bolivia were the first people to cultivate potatoes, possibly 10,000 years ago! This is the oldest cultivated vegetable that I've come across in my research. Potatoes were so well respected and appreciated by the Incans that they represented potatoes in their ceramics and food-storage vessels. This included depictions of the proto-Mr. Potato Head, a human-potato hybrid. Weird! Potatoes grown on the Andean Plateau served as the major energy source for the Incans. They were prepared in numerous ways, including baking and mashing as we do today. Incans also repeatedly freeze-thawed their potatoes to make them softer and more palatable, called chuño. Neat idea! These potatoes were then dried to increase the shelf life and included in stews. Potatoes don't preserve well, so the earliest archaeological evidence is only about 4,000 years old, discovered in Ancón, Peru.

Potatoes and corn made their way to Europe from South America, carried back by Spanish sailors in the late 1500s as a food source during the long voyage home. Uneaten tubers were planted. It is believed that Sir Walter Raleigh was the first European to grow potatoes in Ireland in 1589. Europeans in South America first shunned potatoes, viewing them as a food for the natives only. In Europe, potatoes were thought to be a food for the working class and avoided by the upper class well into the 1600s. Potatoes have since become a staple food for people of nearly all races and classes. Case in point, the introduction of potato agriculture to Ireland is believed to have prevented a Malthusian trap (when a species population becomes too large to be supported by the available food) and enabled the Irish population to grow into the millions.

Perhaps the clearest example of the importance of potatoes for subsistence is the famous potato famine in Ireland that began in 1845 and ended in 1849. This famine was caused by a fungal growth, Phytophthora, known as "late blight" or "potato blight," which destroys the stems, leaves, and causes the tubers to rot in the ground. While cereal grains weren't too difficult to grow, even in western Ireland farmers turned to potatoes in the early 1800s. The potato plant thrived in the Ireland climate, and the tuber was energy and nutrient dense. As a result, approximately 50 percent of rural tenants became nearly completely dependent on potatoes for survival. Unwittingly, Irish farmers hurt themselves by growing only the two most robust types of potato plants, thus reducing the genetic variability in their crops and making entire potato crops susceptible to disease. Then, potato blight arrived from North America, all of the Irish potato plants died, and all heck broke loose.

The problem was exacerbated because the food that continued to be produced in Ireland was too expensive for locals to buy and was exported. Also, there was no unemployment compensation or welfare for farmers or laborers, and insufficient efforts were made by the British to part with their own grain stores or to import a greater amount of cheaper grains (from the United States) to Ireland to prevent starvation. The Irish people, completely subservient to British landlords, showed no revolt or uprising. Instead, the Irish population dropped by approximately 25 percent, two million people, in just seven years, with an estimated half of these unfortunate souls dying from starvation. There were longer-term impacts as well because, when it achieved independence in 1921, the Irish population was still reduced to half of what it was in the early 1840s.

Potatoes like cool, moist soil, not dry dirt or stone

In Peru alone, there are now over 2,500 varieties of potatoes. However, growers are running out of suitable ground for potato cultivation due to warming temperatures. Potatoes that were grown at elevations 2,800 to 3,200 feet above sea level thirty years ago do not grow there anymore and must be cultivated above 4,000 feet above sea level (Lino Loaiza, Association for Nature and Sustainable Development, ANDES). The higher one moves to grow crops, the more fertile soil is replaced by infertile rock. There is a limit to applying altitude adjustments to offset warmer temperatures.

Farmers in Peru are now working with scientists to study the impacts of increased temperature, disease and insect susceptibility, and more violent precipitation on potato crops and to identify varieties that show propensity for adaptation. In Lima, the International Potato Center (IPC) has created a potato gene bank and has cloned about 70 percent of the world's potato plants. This group also sends potato plants of differing varieties to farmers internationally to have them grow the plants under varying conditions and to report the results back to the IPC. Great idea! This effort will allow for the tailoring of potato cultivation across the globe. Without outside intervention, 25 percent of potato species will become extinct by 2050. Beyond that, no potato chips, no French fries, no baked potatoes, no roasted potatoes.

DELECTABLE POTATO DISHES

Baked potatoes, twice-baked potatoes, loaded baked potatoes, mashed potatoes, roasted potatoes, potato salad, potatoes au gratin, pierogi, potato latkes, potato soup, French fries

Standard Recipe: Potatoes au gratin (Preparation time, 10-15 minutes; Cooking Time, 60 minutes)
   Ingredients:
   4 medium-sized Russet potatoes
   2 tsp salt
   ¼ tsp ground black pepper
   1 cup grated parmesan-romano cheese
   2 ¼ cups heavy cream
   1/8 stick of butter

Directions:
1. Preheat oven to 250° Fahrenheit with oven rack in middle position.
2. Butter an 8-inch baking dish.
3. Peel the potatoes and slice as thinly as possible.
4. Toss the potato slices with the salt and pepper in a large mixing bowl.
5. Arrange a layer of overlapping potato slices on the bottom of the baking dish.
6. Pour a fourth of the cream and sprinkle a fourth of the cheese over this bottom potato layer.
7. Repeat this process until all of the potato slices are used, and pour leftover cream on top.

8. Bake at 250° for 60 minutes uncovered. The potatoes should be pierced easily with a knife, and the top should be a golden brown.

Nonstandard Recipe: Potato and chickpea red curry (Preparation time, 10 minutes; Cooking Time, 45 minutes)

Ingredients:

One 13.6-oz can of full-fat coconut milk

4 oz of red curry paste

1 medium potato cut into chunks

One 13.6-oz can of chickpeas drained and rinsed

1 cup of onions diced

1 tbsp garlic minced

1 tbsp brown sugar

1 tbsp avocado oil

Directions:

1. Pour the avocado oil into a medium pot over medium heat.
2. Put the onion and garlic into the pot and stir until the onion becomes translucent.
3. Put the potato and chickpeas into the pot and stir for 4 minutes.
4. Pour in the coconut milk and brown sugar and stir well.
5. Allow for the pot to come to a boil, then reduce to a simmer for 30 minutes or until the potatoes become tender.

Wine may be considered the Belgian chocolate of alcoholic beverages. In moderate amounts, drinking wine can promote better health, provide psychological comfort, and perhaps increase longevity. With respect to physical health, wine contains both antioxidants and anti-inflammatories. As in other fruits and vegetables described in this book, wine has relatively high levels of tannins (polyphenol antioxidants) and a small amount of a molecule called resveratrol (antioxidant and anti-inflammatory), which is part of the plant's immune system.

CHAPTER 15

○ ○○○○○ ○○○○○○ ○ ○ ○○○○ ○○○○○○

How Will I End My Day Without a Glass of Wine?

The efficacy of resveratrol as an anti-inflammatory agent in humans is still up for debate, and pharmaceutical companies have struggled to make a form of resveratrol that shows substantial bioactivity. Antioxidants and anti-inflammatories are also good for the brain and can improve mental health and cognitive ability. However, alcohol is bad for the brain in just about any amount when ingested chronically, so there are better sources for these healthy molecules.

Wine is paleolithic . . .
pre-agriculture, pre-economy, pre-society

Grape fermentation was born at least 10,000 years ago in what is considered the cradle of wine production situated between the Black and Caspian Seas. At that time, this land was called the South Caucuses. Now, it is the nation of Georgia. They squeezed the juice from grapes into pots and then buried these pots called *qvevri* underground for the winter. In the spring, they had wine. Some of these qvevri remained underground for decades, producing a finer, aged wine. Pottery shards found in Jiahu in central China date grape fermentation in that region back about 9,000 years. Biomolecular archaeologist Patrick McGovern and his team have also discovered evidence of grape fermentation residues (tartaric acid) in 7,000-year-old jars discovered in the Zagros Mountains of Iran. The presence of the wine preservative terebinth tree resin confirmed that the grapes were intentionally fermented to produce wine.

The first known winery, containing a grape press and fermentation vats, was discovered in Armenian caves and dates back about 6,000 years. Around the same time, the ancient Greeks and Romans worshipped a god of wine (and ecstasy) called Dionysus or Bacchus, and the ancient Egyptians worshipped Osiris, Lord of the Underworld, whose blood was the source of wine. Perhaps better known in the US are the stories from the Bible describing Noah as the originator of wine (who got plastered and passed out during his first wine experience) and depicting the Wedding at Cana, where Jesus turned water into wine. Across time and space, wine has been a religious symbol and a status symbol representing happiness, friendship, sustenance, and transformation.

Modern wine history is only 400 years old

The grapevine constitutes the genus *Vitis*, which sits in the family of flowering plants known as *Vitaceae*. Appropriately, grape growers are referred to as viticulturalists or vintners. There are seventy-nine defined species of grapevines and thousands of cultivars. About three-quarters of commercially produced grapes are used to make wine. The grapevine that is grown in European and American vineyards is predominantly *Vitis vinefera*, which originated in Europe and began to be imported to North America early in the sixteenth century. However, the first successful cultivation of European grapevines in North America was in the drier, sandier soils of New Mexico in 1629, where blight and the fungal disease "black rot" could be avoided.

While the thought of transferring pests between continents was not considered in the early years of modern-day wine production, it also was not a problem until the invention of the steamship, which reduced travel time to a shortened span that could be survived by more bugs. The result was the Great French Wine Blight of the mid-nineteenth century. The culprits were *Phylloxera* aphids, thought to have arrived in Europe in the mid-1850s as stowaways on ships coming from North America. Over 40 percent of French vineyards were destroyed for a fifteen-year period, from the late 1850s to the early 1870s. Many traditional and some not so traditional approaches to solve the problem were attempted, including pesticides, toads, and free-range poultry. None of these approaches worked. American grapevines were imported, but only a few were shown to grow successfully and produce grapes in the French vineyard soil that was high in calcium bicarbonate content (i.e. "chalky" soil). It wasn't until the aphid-resistant American grapevine *Vitis riparia* was grafted with the *Vitis vinifera* vine of Europe in the late 1880s that the French wine industry started to recover.

When we think "good wine," we think "location, which translates to "environment"

Perhaps no fruit plant is linked to climate more strongly

than the grapevine. Vineyards tout the microclimates that give their wines that special edge. Leading wine makers will tell you that growing the grapes is the most important part of the wine-making process, more important than anything that happens to the grape in the cellar after it is picked. Even the most traditional wine producers will tell you that there are four factors that make a wine: weather, soil, topography, and variety of grape. Simply put, the grapes make the wine, and the climate makes the grapes.

Across the globe, climate zones that support wineries vary in average temperature by only eighteen degrees Fahrenheit. For some types of wines, the range is even narrower. As such, the pinot grape with a microclimate range of only two degrees Celsius may be the first major casualty of climate change. Region-wise, most of Southern and Western Europe will become totally unreceptive to grapevines. However, across the last few decades, areas that were once inhospitable to grapevines have warmed to become inviting microclimates. For instance, places in southern England that were once too cold for vineyards, like Essex, Kent, and Sussex, now experience warmer and drier summers that are quite nurturing to grapevines. Similar stories can be told about regions in southern Germany and northern France. This may sound like good news, and it is. But these regions are transitioning from being too cold to becoming too warm for grapevines and are currently at a temporary sweet spot that likely will not last beyond this century. Even these regions will eventually become hostile to grapevines. This brief description does not include effects of rises in ocean waters in reducing the land area available for wine production and inundation by insects, once limited to lower latitudes, that will become able to survive at higher wine-growing latitudes.

Renowned viticulturalist Richard Smart has referred to wine as the canary in the coal mine of global warming. The first thing to go will be wine quality, which is heavily impacted by growing temperature. Higher temperature equates to faster grape maturation, resulting in higher sugar content and lower acidity levels when picked. These heartier grapes will produce more alcohol when fermented and will be sweeter to the taste, with bolder flavors, including the taste of alcohol. No more dry wines. No more delicate wines. In fact, the E. and J. Gallo Winery in Modesto, California, has already rebranded Thunderbird, which was a citrus-tasting, sweet wine sold to the general public in the 1950s to 1960s. However, it will now cost far more than sixty cents (referring to a radio jingle from the late 1950s, "What's the word? Thunderbird! What's the price? Thirty twice!"). In the longer term, there simply will not be many microclimates that support wine vineyards. At the current rate of warming, we can expect a greater than 80 percent decrease in wine production by the end of this century.

Can the global wine industry save itself?

Efforts to protect wine from climate change are at least fifteen years old. The First World Conference on Climate Change and Wine in Barcelona in 2006 was organized by Pancho Campo, a former world-class tennis athlete and coach and the first Spanish master of wine. This first conference was a media event mostly, and no policies or advances in wine production resulted. Subsequent conferences included keynote speakers like Al Gore, Kofi Annan, and Barack Obama, eventually leading to the Porto Protocol in 2018, which established principles and measures to adapt the wine industry to climate change. Since then, discussions of the problem and creation of abatement plans have dominated the winery headlines. More recently, Michelle Mozell and Liz Thach of the Sonoma State University Wine Business Institute published lists of practical changes to reduce the impacts of climate change on wine production, one for vineyard adaptations and one for cellar production operations. These experts suggest activities to maintain the soil-water balance, prevent soil erosion, boost harvesting, conserve water, reduce CO_2 and methane production, and optimize fermentation.

DELECTABLE WINE DISHES

Straight out of the bottle, mulled wine, red wine spritzer, linguini with wine and clams, red wine stew, mussels in white wine sauce, meats in a white wine sauce

Standard Recipe: Straight from the bottle (Breathing time, 10–15 seconds; Cooking Time, 0 minutes)

Ingredients:

Your favorite type of wine

Directions:

1. Remove cork
2. Pour into your favorite glass

Nonstandard Recipe: Linguini with Clams in White Wine Sauce (Preparation time, 5 minutes; Cooking Time, 15–20 minutes)

Ingredients:

1 lb of linguini (we used spaghetti for this dish)

1 tbsp of garlic minced

3 tbsp of olive oil

1–1 ½ cups of kalamata olives

½–¾ cup of white wine

1 lb of clams

Directions:

1. Clean the clams using a toothbrush and tap water.
2. Cook the linguini al dente in a large pot of boiling water with a pinch of salt, then drain.

3. Pour the olive oil in a pan over medium heat, then sauté the minced garlic for 3–5 minutes.
4. Add the kalamata olives and sauté for another 5 minutes.
5. Add the wine and let the sauce reduce in half.
6. Add the clams and close the lid, then cook for 3 minutes.
7. Add the linguini and toss to coat with the sauce.
8. Remove from the heat and serve immediately.

CHAPTER 16

∘ ∘∘∘∘∘∘∘∘∘∘∘∘∘∘ ∘∘∘∘∘∘ ∘∘∘∘∘∘∘

Saving the Foods We Love

Okay, I know those of us who start every day with a mug of medium French roasted coffee or end every week or every day with a glass of pinot noir may be freaking out right about now. When talking about this book with health care audiences prior to publication, the loss of avocados seemed to prompt a strong emotional response in many attendees. Personally, for me, losing coffee and chocolate is a really big deal. So many delicious and nutritious foods are threatened, but the cause is not lost. This situation can be turned around, and many of the solutions are easy and can be accomplished by anyone, but need to be addressed by everyone.

Human consumption is beyond this world

Without getting too philosophical about the meaning of happiness or too deep into cognitive reframing, it is pretty obvious that many things that we think we need we really just want or like and probably can do without. When surrounded by gluttony, hoarding, and excess, it is sometimes easy to think, *Well, I'm not as bad as that guy.* When surrounded by emaciation, poverty, and desperation, things look different. On top of this, some of our everyday behaviors are actually really unhealthy, like driving two or three blocks to get some fast food. Or continually buying single-serving water bottles instead of a home water filter and a few reusable bottles, which is actually less expensive in the long run. Or using an escalator or an elevator instead of stairs to ascend or descend only one or two floors, even though one is healthy and unencumbered or has a goal of losing a few pounds. Or pressing electric door-opening buttons that are meant for handicapped individuals in wheelchairs. The list is very long. Giving up these bad habits is win-win for the individual and for the environment.

We are the only species on this planet that can mitigate global warming. We may get some assistance from bears as they poop out cherry pits at higher elevations, but we can't rely on pooping bears to solve this problem. We have specialized structures in the frontal lobes of our massive brains that allow us to do some cognitive things that other animals simply don't do anywhere near as well as we do. One of these traits is planning for the distant future. More than any other animal on this planet, we drop money into our 401K, 403B, and IRA retirement plans. Nearly all of us save money for the future, when we can work no longer. We all have this skill set. We just need to apply it in broader ways, not just to money, but to food, energy, air, water, etc.

In the same vein, the concept of delay discounting or future devaluation should be taught in core psychology courses beginning in junior high school or high school. Delay discounting is the propensity to choose a small immediate gratification over a much larger reward that won't arrive for a while. There is a very large body of hybrid economics/psychology/neuroscience literature that describes this fascinating yet depressing phenomenon, and we have a few prominent scientists studying this at Mason. If you want to know more, you can contact Dr. Frank Krueger in the School of Systems Biology, and he will give you a very colorful explanation.

There is less science showing how we can increase future valuation. Why do we devalue the future of our planet and instead overvalue unnecessary merchandise purchases, more device time, and immediate gratification? Dr. Robert Dvorak at the University of South Dakota has shown that an important contributor to delay discounting is inadequate nutrition. In one very interesting study, he asked the students in one of his classes to return to the following class without eating lunch so that everyone was fasted, with low blood sugar, upon arrival. Then, half of the students drank soda with real sugar and the other half drank soda with an artificial sweetener. After a short waiting period for metabolic processes to take place, all of the students completed the same questionnaire. They could either choose a small but immediate gratification or a much larger reward that would not arrive for a substantial amount of time. The students who drank the soda with artificial sweetener tended to choose the immediate gratification, while those who consumed real sugar tended to choose the larger but delayed reward. All of the students had low blood sugar levels when they arrived to class. Those that drank the soda with artificial sweetener remained metabolically unsatisfied and succumbed to immediate gratification. Thus, when metabolism is satisfied, we are more inclined to make choices that benefit us in the distant future. So, by saving these foods from extinction and in turn maintaining adequate nutrition, we satisfy part of the equation to increasing future valuation.

Use less fossil fuel energy

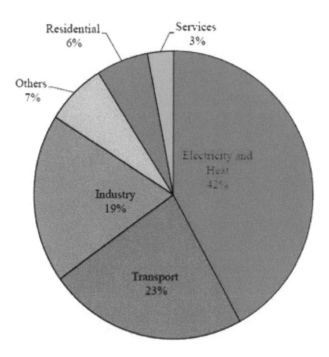

This is the crux of the problem. We burn too much fossil fuel. The pie chart on this page was originally created by Dr. Sultan Al-Salem at the Kuwait Institute for Scientific Research. It shows the distribution of the commercial, industrial, and residential contributions to CO_2 emissions across the globe. Residential sources of CO_2 emissions in the US are double what is seen globally (12 percent, Environmental Protection Agency, 2018). Loosely translated, American homes use way more energy than the global average. We need to reduce our energy consumption and quickly wean ourselves off of fossil fuel combustion in particular. Note that electricity is created in large part by fossil fuel combustion. Simple steps include converting to energy-efficient light bulbs and LEDs (light emitting diodes), adding more insulation to our attics, better sealing attic and cellar joints and around exterior doors and windows (replacing old windows with energy-efficient windows, if the cash flow permits), and, if we can afford it, upgrading home appliances, water heaters, and HVAC systems. We can consider installing solar panels. A typical house can be solarized for about $20,000, which is earned back over the following decade in energy cost reductions and through paybacks by utility companies when excess electricity produced by the home panels is fed back into commercial grids. Home solar panels will last about 30 years before their energy production begins to decline and will continue to provide energy well after this industry marker has been reached.

We can also replace gas-guzzling cars with certified pre-owned hybrid and electric vehicles. Buying new cars is probably not a good idea because the carbon footprint to produce most new cars is very high, requiring more than a lifetime of use to offset. Acknowledged, this last option is not so simple. If we can't immediately replace our inefficient vehicles, can we find ways to drive less? Every gallon of gas we don't burn keeps twenty pounds of CO_2 out of the atmosphere. Given that there are about 300 million vehicles on the US roadways, if every driver could save one gallon of gas per week, this would amount to a reduction of 3 million tons of CO_2 released each week, or 156 million tons annually. To provide more perspective, a standard sedan carries about nineteen gallons of gas and an SUV about twenty-six gallons. So, we're talking about burning only 4 to 5 percent less of a tank of gas per week. In the bigger picture, this is a relatively small change in CO_2 emissions, about 2.5 percent of the annual increase in CO_2 emissions, which is six billion tons. However, the percent reduction goes up even more if we drive even less, and fossil fuel companies are compelled by demand, so the less we buy, the less they produce. This is only one part of a multidimensional effort at reducing reliance on fossil fuels.

We have devices to do everything for us, and we are rapidly evolving into the animated characters from the Walt Disney movie *Wall-E*. Nearly all of these devices use energy in the form of electricity. We don't harvest and store natural electricity from lightning storms. Electricity is generated in large part by fossil fuel combustion, including natural gas. Electrical power outlets are everywhere and free to use in many locations. Airports freely charge phones and tablets for all travelers. Private and public employees charge their personal devices at work. Because of this, it is easy not to think about how much electricity we use on a daily basis. How frequently do we really need to look at our cellular phones? For those of us older than forty, how many times did we look at our cellular phones twenty-five years ago? The answer is close to zero because cell phone use did not become popular until the mid to late 1990s. We were happy (perhaps happier) and just as productive. Is more time on social networking, gaming, and video apps really worth struggling for basic needs and comfort in the future? Can we who know a life without handheld devices advocate a less electronic life to those who might listen?

Buy less stuff, recycle, and produce less waste

We often buy stuff we don't really need because the act of purchasing something new is exciting and makes us feel good. We tend to buy with abandon and throw away rather than to buy with caution, use with care, and maintain. We buy without much forethought. Sure, we price shop a bit and try to tailor most purchases to our personal tastes, but seldom do we ask how the product was made, what will happen to it when we're done using it, and what its total environmental cost is. The more stuff we buy, the more garbage we produce and the more pollution we create. Plastic micro-particles can now be found deep in our oceans, and at mountaintops, and in the bellies, tissues, and cells of numerous land and water-dwelling animals, including sea turtles and seabirds. And we throw away about three billion batteries or 180,000 tons of batteries each year. These are huge problems.

Before buying something, even a staple life item, it is one's responsibility to learn a little bit about the environmental cost of its production, estimate how many times or for how long it will be used, and understand how much pollution it produces while it is being used. And if after all of this deliberation it still seems like a wise purchase, it remains one's responsibility to sell, donate, or recycle the item when it is no longer of personal use and not throw it in the trash. All of these factors contribute to the total carbon footprint of the item, and so what might seem to cost only a few bucks at the register during an impulse purchase in reality costs a lot more to the environment over time. Once a symbol of wealth, the act of under-using and discarding has become engrained in nearly all modern cultures across the planet. This behavior is propagated by formal systems that are maintained by private businesses and public services that, for only a few extra pennies, greatly facilitate our ability to obtain and discard stuff. It is easy to discard because it is easy to discard. If we had to pay more for refuse services or lug our own garbage to a local dump, I'm pretty sure many of us would purchase and discard less.

Does this purchasing responsibility plan seem like a lot of work? If so, what does that say about our willingness to understand and appreciate the world in which we live and do the things that will preserve it? If we value stuff more than the actual medium in which we live, we have lost touch with our true purpose. As Marie Kondo would advise (*The Life-Changing Magic of Tidying Up*, 2011), we should look at our stuff and ask ourselves, "Which of these items really sparks joy?" Then, get rid of everything else.

Don't throw it away! Sell it, donate it, or recycle it. If every item we see on the internet sparks joy, then we are a victims of materialism, which can be cured by a hobby, regular social interaction in a natural setting, or engaging in sports, hiking, camping, or just about anything outdoors. Houses are really just big boxes that require deliberate introduction of various forms of enrichment, like art pieces, paintings, sound systems, video screens, video games, handheld devices, and toys for toddlers. We typically take our girls outside empty handed because the outdoors really requires no additional man-made enrichment. Full disclosure: We introduced man-made enrichment in our backyard by installing a small playground to increase time spent outdoors and promote physical activity and coordination, but more time is spent finding wild strawberries and "candy lands" (dandelions) in the grass and using them to cook up imaginary delicacies.

Of the 650 billion dollars' worth of food that is purchased each year, 165 billion dollars' worth of food is thrown away. This means that, on average, Americans throw away about a quarter of the food that we purchase. Estimates from the US Department of Agriculture put this number closer to 30 percent food waste. Discarded food is the largest component in most municipal landfills. Perhaps not surprising, those who eat healthier diets composed of more fresh fruits and vegetables are the most wasteful (they tend to have disposable incomes, and the food spoils relatively quickly). As all of this wasted food rots and decays, it contributes about 10 percent to the total US methane emissions. Leftovers can simply be reheated or can be transformed into other delicious dishes. If we can't eat some leftover food, it can be composted. As a general rule, if it can be grown or eaten, it can be composted. The compost can then be put back into our gardens to improve health and yield.

**Plant a tree or a garden
and show more appreciation for native plants**

Planting a tree is a one-day, thirty to one-hundred-dollar effort followed by some infrequent watering until roots have fully developed. Don't make the mistakes I did with our miniature apple tree. If you plant a fruit tree, first be sure to read up on what it will need to thrive and bear fruit.

Planting and maintaining a garden is really not that hard. The photo on this page shows our garden as the sun began to set on a mid-July day in 2019. We purchase manure, peat moss, and garden soil on a yearly basis. We add vermiculite or perlite to enhance water retention, but a large bag of that stuff lasts a few years if you limit it to the base of the plant rather than mixing it into the entire bed.

We buy young plants and also grow from seed. We invest about 300 dollars each year to start the garden. We spend one day to make all of the materials purchases and prepare the garden, and one day to purchase and plant the seeds and seedlings. The rest is basically just watering, picking off leaves if they brown or wilt, and tying a few pepper plants to stakes (large or many peppers will pull the plant to the ground). We devote the most time to our tomato plants, stringing them up when they get too tall to support their own weight and pinching off suckers so that they consist of only three to five main branches. The tomato plants shown will reach the top of the deck by the first week of August, and these plants provide all the tomatoes we can eat from late June through late September. We recover all of the money we invest in the garden from the organic vegetable yield. In the spring after each growing season, we take the raised bed soil and use it to maintain our yard, so the soil gets used twice.

The image of the front lawn as a uniform emerald carpet is antiquated and unstainable in a vast majority of American neighborhoods. Lawns consisting of only one grass type are not healthy and require more external maintenance, just as the food plants that have lost genetic diversity are most endangered. Nature does not share the same organization as the indoors, where only some stuff goes here and other stuff goes there. Nature is more complicated than that. Likewise, a healthy yard is not necessarily a function of what appeals to the human eye. Lawns are primarily intended to control ground erosion around the house and should probably consist of grasses or plants that are native to the region, in line with the doctrines of the permaculture movement that endorses the creation of residential environments that are self-sustaining and systematically integrated into local ecosystems. Lawns are primarily intended to control ground erosion around the house and should probably consist of grasses or plants that are native to the region and benefit the local ecosystem.

Native plants have become the homeless mutts of most neighborhoods, overlooked by homeowners who would prefer the purebred in the showcase window. Lawn and garden health can be achieved easily without applying chemicals, if you're okay with some variation in the visual appearance. Moreover, herbicides and pesticides contain chemicals that are suspected to cause human lymphoma and hypothyroidism and are known to negatively affect off-target plants and animals, and they are not contained to the sites to which they are applied. Wind blows these chemical sprays tens to hundreds of meters as they are being applied, and rainwater carries residual unabsorbed sprayed or pelleted chemicals off the property after application. Numerous chemicals that are contained in some residential pesticides and herbicides maim and kills bees. Examples are glyphosate and neonicotinoids. Some herbicide chemicals, like glyphosate and dicamba, also reduce butterfly populations because they kill milkweed plants, a major food source for most butterflies, and host plant for the monarch butterfly. If your soil is a bit acidic and you're concerned that your lawn isn't as full as it might be, consider applying some lime to neutralize the soil instead of a synthetic fertilizer containing herbicide. Neutral pH gives grasses a competitive advantage over mosses.

Buy from local farmers

Farmer's markets are awesome! There is really no better way to start weekend mornings than visiting a local farmer's market as a family. There are free samples everywhere: peaches, apples, plums, melons, tomatoes, cheese, tea, wine, etc. This stuff is perishable, so bring a reusable bag and buy only what you will eat in a week to minimize food waste.

Produce purchased at farmer's markets typically does not travel hundreds of miles in a refrigerated truck burning diesel fuel, and local farmers tend to employ more sustainable agricultural practices, so the carbon footprint of this food is lower than what is purchased in grocery stores and supermarkets. To learn more about this, I refer you back to the three books about US industrial agriculture mentioned in the introduction. Also, farmer's markets sell organic produce that has not been tainted by pesticides and herbicides. Finally, the produce is fresh. It has not been waxed or treated in other ways to preserve color and appearance while it oxidizes and loses nutrients sitting on a shelf somewhere. We literally have to drag our daughters away from the fruit samples at our local farmer's markets.

Boycott businesses with bad practices

The bottom line for any business is profit. If we want to influence a company, we need to mess with their profits. This includes state government profits. One of the easiest ways to do this simply is to not buy stuff from companies that don't act in ways that mitigate their own carbon footprints, and to reward companies that are doing their part to reduce global warming by selectively buying their products. Boycotting works! We've recently seen Kimberly-Clark, Nestle, Nordstrom, Staples, Uber, SeaWorld, and the state of North Carolina impacted by public activism and boycotting.

Engage politicians

There are arguments that this massive climate change problem is too cumbersome to be dealt with directly by citizens at large and cannot be solved without some sort of governmental and/or industrial intervention. This may or may not be true. It is certainly true that changes at the industrial level can relatively easily turn the tide on this problem. Obvious examples were revealed while this book was being prepared and during the coronavirus (Covid-19) pandemic of 2019–2020 that swept the globe over the course of three months.

The first example was an astounding series of satellite images (taken by the European Space Agency) of changes in greenhouse gas emissions in China after all unnecessary manufacturing was suspended in an attempt to contain the virus. Over the course of just a few weeks, a massive reduction in nitrogen dioxide emission was recorded. A decrease in carbon emissions of more than 20 percent was also recorded at the peak of the industry shutdown. Soon after, the same effect was viewed by satellites targeting Italy, when the government shut down all nonessential businesses and quarantined its citizens. Similar effects were also observed in the US after Governor Andrew Cuomo declared a disaster emergency in the state of New York in early March 2020. Carbon dioxide, carbon monoxide, and methane levels in New York City dropped by 10 percent in less than a month. Combined, these industrial changes did not prevent a rise in carbon emissions, but they slowed the increase. Unfortunately, if the greenhouse gas emission pattern during the coronavirus epidemic is anything like that during the economic shutdown during the 2008 global financial crisis, emissions will rebound with a fury that erases any gains made after the most serious phase of the health crisis is over.

The petroleum industry collapsed about two months into the coronavirus pandemic, to the point where oil companies in the US had to pay others to take their product. One might have seen this as an opportunity to begin to do away with fossil fuel and restart a different, greener energy engine for our nation. The timing was right to contact state and federal representatives who witnessed the massive reduction in greenhouse gas emissions and could no longer hide behind a curtain of uncertainty or argue that greenhouse gas reduction simply couldn't be achieved. It happened. The air got cleaner. The smog over major metropolitan areas dissipated. If there was a positive side to the coronavirus pandemic, it was to illustrate that we can all survive when the gears of global retail markets stop turning.

This is the sort of stuff to include in a letter to one's federal representatives urging the transition to more planet-friendly economic systems and the creation and enforcement of regulations that make dirty industries liable for their carbon footprints. Try to find ways to inspire friends and family to sound the alarm. While they may not show any gratitude or allow you a sense of satisfaction, they are likely altering their behavior even just a little bit when you're not looking, and they might just send a letter, so make the effort.

However, we can't expect much of a response from any politician if we don't make some sort of investment in the communication. Not a financial investment, but mental effort to create thoughtful fact-based arguments and propose solutions to the problem (like carbon solidification) and not rely on politicians to find the answers on their own. We can also learn more about our politicians' likes and dislikes and use this information to find inroads for communication and tailor our letters. Finally, we can make our protests more impactful by voting in climate-change activists and voting out climate-change deniers. We need to know our candidates and vote for those that are green, not necessarily blue or red. Ralph Nader, an attorney, consumer advocate, environmental activist, and politician who ran for US president five times between 1992 and 2008 (as the Green Party candidate in 2000), was ahead of his time with respect to public awareness and acceptance of climate change. Sincere gratitude to Mr. Nader for his efforts. Do we need a re-emergence of the Green Party, or maybe just some sort of Green New Deal?

APPENDIX
○○○○○○○○○○○○○○○○○○○○○○○○○○○○○○
Health-Promoting Nutrients, Minerals, and Phytochemicals in Endangered Foods

Nutrients

<u>Vitamin A</u> (retinol) – A potent antioxidant. Important regulator of embryonic and fetal growth. Required for vision. Helps maintain skin and mucous membrane health (mouth, nose, eyelids, trachea, lungs, esophagus, gut), cell differentiation. Beta-carotene is a precursor to retinol found in endangered foods.

* cherries, chickpeas

<u>Biotin</u> – An essential cofactor for numerous enzymatic reactions. Regulator of gene expression. Must be ingested, not produced in the human body.

* Avocados, peanuts

Vitamin B2 (riboflavin) – A major player in catabolism, breaking down carbohydrates, fats, and proteins for energy. Takes part in oxygen utilization.

- avocados

Vitamin B5 (pantothenic acid) – A major player in catabolism, breaking down carbohydrates and fats for energy.

- coffee

Vitamin B6 (pyridoxine) – A major player in catabolism of carbohydrates for energy, and metabolism of proteins. A cofactor in over 100 chemical reactions in the body. Takes part in the production of neurotransmitters, like serotonin and norepinephrine. Required for proper brain development. Helps maintain heart, immune, and muscle function.

- bananas, fish, peanuts, potatoes

Vitamin B9 (folate) – A critical player in metabolism of protein and DNA. Critical in embryonic development. Required for proper postnatal growth in children. Involved in mood regulation.

- avocados

Vitamin C (L-ascorbic acid, citric acid) – A potent antioxidant. Essential cofactor in numerous enzymatic reactions, synthesis of neuropeptides, and gene expression. Helps to lower systemic inflammation. Required for gum health and wound healing (the deficiency is known as scurvy).

- apples, bananas, cherries, honey, potatoes

Vitamin K (menadiol) – A coagulant factor. Involved in calcium metabolism. Regulates bone, circulatory system, and brain health.

- avocados, cherries

Minerals

Calcium – The most abundant mineral in the body. Major component of bones and teeth. Integral factor in cell-signaling pathways. Required for bone metabolism, brain function, heart contraction, and muscle strength. Calcium is the major neuroplasticity ion in the brain.

- chickpeas, grapes

Copper – A cofactor to many enzymes, including those in the energy cycle of every cell. Regulates iron metabolism. Involved in bone, blood vessel, brain, immune, muscle, and nerve health.

- avocados, bananas, cherries, chickpeas, chocolate

Iron – A basic component of hundreds of proteins and enzymes involved in essential cell functions. Involved in blood cell generation, processing of oxygen by hemoglobin, energy metabolism, and regulation of body temperature. Helps maintain immune function.

- beer, chickpeas, chocolate

Magnesium – A cofactor in over 100 chemical reactions in the body, especially energy metabolism. Regulates protein and nucleic acid (DNA, RNA) synthesis. Required for bone formation and calcium absorption into bone cells. Promotes immune, metabolic, and psychological health.

- bananas, chickpeas, chocolate, coffee, potatoes

Manganese – A cofactor in antioxidant processes. Involved in carbohydrate, fat, and amino acid (glutamine) metabolism, blood sugar regulation, and calcium absorption. Helps maintain brain and immune health.

- cherries, chickpeas, chocolate, coffee, potatoes

Phosphorus – A necessary component of cell membranes and nucleic acids (DNA, RNA). Required for bone and tooth formation. Critical for cell-signaling and brain plasticity. Integral to energy metabolism, maintenance of bone density, protein formation, and filtration of waste from the blood in the kidneys.

- chickpeas, chocolate, potatoes

Potassium – A critical electrolyte in maintaining basic functions of all cells in the body. Necessary for brain function, kidney filtration, heart contraction, muscle strength, and water balance.

- bananas, cherries, chocolate, coffee, potatoes

Selenium – A structural component of proteins. A cofactor in antioxidant and anti-inflammatory processes. Helps regulate physiological responses to stress and maintain brain health, immune function, and fertility in men and women.

- chocolate

Zinc – A player in catabolism, breaking down carbohydrates for energy. Required for proper cell division, cell growth, organism growth, and healing. Helps maintain brain health, gut processes, and immune function.

- chickpeas, chocolate

Phytochemicals

Anthocyanins – Water-soluble plant pigments that act as potent antioxidants.

- bananas

Polyphenols (immune factors in plants) – Potent antioxidants. Various and growing recognized functions, including maintaining brain function, cardiovascular and immune health.

- apples, beer, cherries, chocolate, wine

REFERENCES

○ ○○○○○○ ○○○○○○○ ○ ○○○○○ ○○○○○○○

Introduction

Hyman, Mark (2020) Food Fix. How to Save Our Health, Our Economy, Our Communities, and Our Planet—One Bite at a Time. Little, Brown Spark. New York.

Little, Amanda (2019) The Fate of Food: What We'll Eat in a Bigger, Hotter, and Smarter World. Harmony Books. New York.

Tickell, Josh (2017) Kiss the Ground. Enliven Books (Simon and Schuster, Inc.). New York.

Zhou N, Gu X, Zhuang T, Xu Y, Yang L, Zhou M. (2020) Gut Microbiota: A Pivotal Hub for Polyphenols as Antidepressants. Journal of Agriculture Food Chemistry. Epub ahead of print.

Climate Change

International Plant Protection Convention (IPPC) (2013) Warming of the Climate System is Unequivocal. Report on Climate Change 2013: The Physical Science Basis—Summary for Policymakers, Observed Changes in the Climate System.

NOAA.gov

Ridley, Matt (2003) Nature via Nurture: Genes, Experience, and What Makes Us Human. Harper Collins Publishers. New York.

Solomon, S., J. S. Daniel, T. J. Sanford, et al., 2010: Persistence of climate changes due to a range of greenhouse gases, Proceedings of the National Academy of Science, 107: 18354-18359.

Solomon S., G.-K. Platter, R. Knutti, et al., 2009: Irreversible climate change due to carbon dioxide emissions, Proceedings of the National Academy of Science, 106: 1704-1709.

Venkataramanan, M (2011) Causes and effects of global warming. Ind. J. Sci. Technol. 4(3):226-229.

Apples

Subbaiah TK, Powell LE (1992) Abscisic acid relationships in the chill-related dormancy mechanism in apple seeds. Plant Growth Regulation. 11(2):115-123

Cornille A, Gladieux P, Smulders MJM, Roldán-Ruiz I, Laurens F, Le Cam B, et al. (2012) New Insight into the History of Domesticated Apple: Secondary Contribution of the European Wild Apple to the Genome of Cultivated Varieties. PLoS Genet 8(5): e1002703.

"Apple production in 2017; Crops/World Regions/Production Quantity." Food and Agriculture Organization of the United Nations (fao.org). Archived from the original on 11 May 2017.

Sugiura T, Ogawa H, Fukuda N, Moriguchi T. (2013) Changes in the taste and textural attributes of apples in response to climate change. Sci Rep. 2013;3:2418.

Avocados

Barlow, Connie. The Ghosts of Evolution. Basic Books, 2000.

Collins M, Thrasher A. (2015) Gene therapy: progress and predictions. Proc. Biol. Sci. 282(1821):2014.3003.

Rendón-Anaya M, Ibarra-Laclette E, Méndez-Bravo A, Lan T, Zheng C, Carretero-Paulet L, Perez-Torres CA, Chacón-López A, Hernandez-Guzmán G, Chang TH, Farr KM, Barbazuk WB, Chamala S, Mutwil M, Shivhare D, Alvarez-Ponce D, Mitter N, Hayward A, Fletcher S, Rozas J, Sánchez Gracia A, Kuhn D, Barrientos-Priego AF, Salojärvi J, Librado P, Sankoff D, Herrera-Estrella A, Albert VA, Herrera-Estrella L. (2019) The avocado genome informs deep angiosperm phylogeny, highlights introgressive hybridization, and reveals pathogen-influenced gene space adaptation. Proc Natl Acad Sci U S A. 2019 Aug 20;116(34):17081-17089.

Bananas

Varma V, Bebber DP (2019) Climate change impacts on banana yields around the world. Nature Climate Change. 9:752-757.

Ordonez N, Seidl MF, Waalwijk C, Drenth A, Kilian A, Thomma BPHJ, Ploetz RC, Kema GHJ. (2015) Worse comes to worst: bananas and panama disease—when plant and pathogen clones meet. PLoS Pathog. 11, e1005197.

Queensland Department of Agriculture and Fisheries, South Johnstone, Australia. https://www.daf.qld.gov.au/

German Calberto, G., C. Staver and P. Siles. 2015. An assessment of global banana production and suitability under climate change scenarios, In: Climate change and food systems: global assessments and implications for food security and trade, Aziz Elbehri (editor). Food Agriculture Organization of the United Nations (FAO), Rome, 2015.

Santos AS, Amorim EP, Ferreira CF, Pirovani CP. (2018) Water stress in Musa spp.: A systematic review. PLoS One. 2018 Dec 3;13(12):e0208052.

Beer

http://www.beer100.com/history/beerhistory.htm, "The History of Beer."

http://blog.drinktec.com/beer/abbey-breweries, "Holy Brews." by Mareike Hassenbeck, August 27, 2018.

Davies M. (2003) The role of GABA$_A$ receptors in mediating the effects of alcohol in the central nervous system. J. Psychiatry Neurosci. 28(4):263-74.

Xie W, Xiong W, Pan J, Ali T, Cui Q, Guan D, Meng J, Mueller ND, Lin E, Davis SJ. (2018) Decreases in global beer supply due to extreme drought and heat. Nat. Plants. 4(11):964-973.

Jackson, Michael (1977) World Guide to Beer.

Cherries

https://www.atlasobscura.com/articles/bing-cherry; Anne Ewbank (2019) The Tragic Roots of America's Favorite Cherry, Gastro Obscura.

Livni, Ephrat (2018) Confused by typhoons, Japan's cherry blossoms are blooming in autumn. Quartz Magazine

Penelope F. Measham et al. Climate, Winter Chill, and Decision-making in Sweet Cherry Production. HortScience, March 2014

Primack RB, Higuchi H, Miller-Rushing AJ (2009) The impact of climate change on cherry trees and other species in Japan. Biol. Conservation. 142:1943-1949

"Will climate change kill the Michigan cherry?" Glen Arbor Sun, Jacob Wheeler, Editor, July, 25, 2012.

Naoe S, Tayasu I, Sakai Y, Masaki T, Kobayashi K, Nakajima A, Sato Y, Yamazaki K, Kiyokawa H, Koike S. (2016) Mountain-climbing bears protect cherry species from global warming through vertical seed dispersal. Curr Biol. 2016 Apr 25;26(8):R315-6.

Naoko Abe (2019) Sakura Obsession. Penguin Random House, LLC, Alfred A. Knopf Publishing. New York.

Chickpeas

Jain M, Misra G, Patel RK, Priya P, Jhanwar S, Khan AW, Shah N, Singh VK, Garg R, Jeena G, Yadav M, Kant C, Sharma P, Yadav G, Bhatia S, Tyagi AK, Chattopadhyay D. (2013) A draft genome sequence of the pulse crop chickpea (Cicer arietinum L.). Plant J. 74(5):715-29.

Singh J, Singh V, Sharma PC. (2018) Elucidating the role of osmotic, ionic and major salt responsive transcript components towards salinity tolerance in contrasting chickpea (Cicer arietinum L.) genotypes. Physiol Mol Biol Plants. 24(3):441-453.

Narnoliya L, Basu U, Bajaj D, Malik N, Thakro V, Daware A, Sharma A, Tripathi S, Hegde VS, Upadhyaya HD, Singh AK, Tyagi AK, Parida SK. (2019) Transcriptional signatures modulating shoot apical meristem morphometric and plant architectural traits enhance yield and productivity in chickpea. Plant J. 98(5):864-883.

Varshney RK, Pandey MK, Bohra A, Singh VK, Thudi M, Saxena RK. (2019) Toward the sequence-based breeding in legumes in the post-genome sequencing era. Theor Appl Genet. 132(3):797-816.

von Wettberg EJB, Chang PL, Başdemir F, Carrasquila-Garcia N, Korbu LB, Moenga SM, Bedada G, Greenlon A, Moriuchi KS, Singh V, Cordeiro MA, Noujdina NV, Dinegde KN, Shah Sani SGA, Getahun T, Vance L, Bergmann E, Lindsay D, Mamo BE, Warschefsky EJ, Dacosta-Calheiros E, Marques E, Yilmaz MA, Cakmak A, Rose J, Migneault A, Krieg CP, Saylak S, Temel H, Friesen ML, Siler E, Akhmetov Z, Ozcelik H, Kholova J, Can C, Gaur P, Yildirim M, Sharma H, Vadez V, Tesfaye K, Woldemedhin AF, Tar'an B, Aydogan A, Bukun B, Penmetsa RV, Berger J, Kahraman A, Nuzhdin SV, Cook DR. (2018) Ecology and genomics of an important crop wild relative as a prelude to agricultural innovation. Nat Commun. 9(1):649.

Varshney RK, Thudi M, Roorkiwal M, He W, Upadhyaya HD, Yang W, Bajaj P, Cubry P, Rathore A, Jian J, Doddamani D, Khan AW, Garg V, Chitikineni A, Xu D, Gaur PM, Singh NP, Chaturvedi SK, Nadigatla GVPR, Krishnamurthy L, Dixit GP, Fikre A, Kimurto PK, Sreeman SM, Bharadwaj C, Tripathi S, Wang J, Lee SH, Edwards D, Polavarapu KKB, Penmetsa RV, Crossa J, Nguyen HT, Siddique KHM, Colmer TD, Sutton T, von Wettberg E, Vigouroux Y, Xu X, Liu X. (2019) Resequencing of 429 chickpea accessions from 45 countries provides insights into genome diversity, domestication and agronomic traits. Nat Genet. 51(5):857-864.

Chocolate

www.climate.gov

https://www.climate.gov/news-features/climate-and/climate-chocolate

Brodwin, E. (2017) Chocolate is on track to go extinct in forty years. The Independent, Business Insider, U.K.

Niether W, Smit I, Armengot L, Schneider M, Gerold G, Pawelzik E. (2017) Environmental Growing Conditions in Five Production Systems Induce Stress Response and Affect Chemical Composition of Cocoa (Theobroma cacao L.) Beans. J Agric Food Chem. 65(47):10165-10173.

Gupta RM, Musunuru K. (2014) Expanding the genetic editing tool kit: ZFNs, TALENs, and CRISPR-Cas9. J Clin Invest. 2014 Oct;124(10):4154-61.

Hsu PD, Lander ES, Zhang F. (2014) Development and applications of CRISPR-Cas9 for genome engineering. Cell. 2014 Jun 5;157(6):1262-78.

Coffee

Acheson KJ, Zahorska-Markiewicz B, Pittet P, Anantharaman
K, Jéquier E. (1980) Caffeine and coffee: their influence on
metabolic rate and substrate utilization in normal weight and
obese individuals. Am J Clin Nutr. 33(5):989-97.

Aime MC, Phillips-Mora W. (2005) The causal agents of witches'
broom and frosty pod rot of cacao (chocolate, Theobroma
cacao) form a new lineage of Marasmiaceae. Mycologia.
97(5):1012-22.

Loftfield E, Freedman ND. (2016) Higher coffee consumption
is associated with lower risk of all-cause and cause-specific
mortality in three large prospective cohorts. Evid Based Med.
21(3):108.

Fish

Hare JA, Morrison WE, Nelson MW, Stachura MM, Teeters EJ, Griffis RB, Alexander MA, Scott JD, Alade L, Bell RJ, Chute AS, Curti KL, Curtis TH, Kircheis D, Kocik JF, Lucey SM, McCandless CT, Milke LM, Richardson DE, Robillard E, Walsh HJ, McManus MC, Marancik KE, Griswold CA. (2016) A vulnerability assessment of fish and invertebrates to climate change on the northeast US continental shelf. PLoS One. 11(2):e0146756.

Innes JK, Calder PC. (2018) Omega-6 fatty acids and inflammation. Prostaglandins Leukot Essent Fatty Acids. 132:41-48.

Worm B, Barbier EB, Beaumont N, Duffy JE, Folke C, Halpern BS, Jackson JB, Lotze HK, Micheli F, Palumbi SR, Sala E, Selkoe KA, Stachowicz JJ, Watson R. (2006) Impacts of biodiversity loss on ocean ecosystem services. Science. 314(5800):787-90.

Petitgas P, Alheit J, Peck MA, Raab K, Irigoien X, Huret M, et al. 2012. Anchovy population expansion in the North Sea. Mar Ecol Prog Ser. 2012; 444: 1-13.

Honey

Egan PA, Adler LS, Irwin RE, Farrell IW, Palmer-Young EC, Stevenson PC. (2018) Crop Domestication Alters Floral Reward Chemistry With Potential Consequences for Pollinator Health. Front Plant Sci. 9:1357.

Gezon ZJ, Inouye DW, Irwin RE. (2016) Phenological change in a spring ephemeral: implications for pollination and plant reproduction. Glob Chang Biol. 22(5):1779-93.

Kerr JT, Pindar A, Galpern P, Packer L, Potts SG, Roberts SM, Rasmont P, Schweiger O, Colla SR, Richardson LL, Wagner DL, Gall LF, Sikes DS, Pantoja A. (2015) CLIMATE CHANGE. Climate change impacts on bumblebees converge across continents. Science. 349(6244):177-80.

The Ladies of the Philoptochos Society (1984) Let's Cook Greek. Irene Emmanuel & Martha Konefal Eds. St. George's Greek Orthodox Church. Hartford, Ct.

Ogilvie JE, Griffin SR, Gezon ZJ, Inouye BD, Underwood N, Inouye DW, Irwin RE. (2017) Interannual bumble bee abundance is driven by indirect climate effects on floral resource phenology. Ecol Lett. 20(12):1507-1515.

Soroye P, Newbold T, Kerr J. (2020) Climate change contributes to widespread declines among bumble bees across continents. Science. 367(6478):685-688.

Peanuts

Bagley, M. (2017) George Washington Carver: Biography, inventions & quotes. Livescience.com. December 07, 2013.

Demby, G (2014) George Washington Carver, the Black History Montheist of them all. NPR.org. February, 11, 2014.

Lovelock, J (2009) James Lovelock on biochar: Let the earth remove CO2 for us. Theguardian.com. March 24, 2009.

Milman, O (2018) Ex-NASA Scientist: 30 years on, world is failing "miserably" to address climate change. Theguardian.com. June 19, 2018.

Tan G, Wang H, Xu N, Junaid M, Liu H, Zhai L. (2019) Effects of biochar application with fertilizer on soil microbial biomass and greenhouse gas emissions in a peanut cropping system. Environ. Technol. 28:1-11.

Potatoes

Ames, Mercedes; Spooner, David (2008). "DNA from herbarium specimens settles a controversy about origins of the European potato". American Journal of Botany. 95(2):252-257.

Joel Moykr, Britannica, 2019

Salaman, Redcliffe N; W. G Burton; J. G Hawkes (1985). The history and social influence of the potato. Cambridge; New York: Cambridge University Press.

Semla M, Goc Z, Martiniaková M, Omelka R, Formicki G. (2017) Acrylamide: a common food toxin related to physiological functions and health. Physiol Res. 2017 May 4;66(2):205-217.

Wine

Bahare Salehi, Abhay Prakash Mishra, Manisha Nigam, Bilge Sener, Mehtap Kilic, Mehdi Sharifi-Rad, Patrick Valere Tsouh Fokou, Natália Martins, Javad Sharifi-Rad (2018) Resveratrol: A Double-Edged Sword in Health Benefits. Biomedicines. 6(3):91.

Hames, Gina (2010). Alcohol in World History. Routledge. p. 17.

Mozell, M, Thach L (2014) The impact of climate change on the global wine industry: Challenges and solutions. Wine Economics and Policy. 3(2):81-19.

Prehistoric China - The Wonders That Were Jiahu The World's Earliest Fermented Beverage. Professor Patrick McGovern the Scientific Director of the Biomolecular Archaeology Project for Cuisine, Fermented Beverages, and Health at the University of Pennsylvania Museum in Philadelphia. Retrieved on 3 January 2017.

Saving the Foods We Love

Alimba CG, Faggio C. (2019) Microplastics in the marine environment: Current trends in environmental pollution and mechanisms of toxicological profile. Environ. Toxicol. Pharmacol. 68:61-74.

Al-Salem S. (2015) Carbon dioxide (CO_2) emission sources in Kuwait from the downstream industry: critical analysis with a current and futuristic view. 81:575e87.

"Our planet is drowning in plastic pollution," United Nations Environment Programme, May 25, 2018.

Benoit RG, Gilbert SJ, Burgess PW. (2011) A neural mechanism mediating the impact of episodic prospection on farsighted decisions. J. Neurosci. 31(18):6771-9.

Bertossi E, Tesini C, Cappelli A, Ciaramelli E. (2016) Ventromedial prefrontal damage causes a pervasive impairment of episodic memory and future thinking. Neuropsychologia. 90:12-24.

Derraik JGB (2002) The pollution of the marine environment by plastic debris: a review. Marine Pollution Bull. 44(9):842-52.

Fischer J, Muller T, Spatz A-K, Greggers U, Grunewald B, Menzel R (2014) Neonicotinoids interfere with specific components of navigation in honeybees. Public Library of Science One. 9(3):e91364.

Friedli A, Williams GR, Bruckner S, Neumann P, Straub L. (2020) The weakest link: Haploid honey bees are more susceptible to neonicotinoid insecticides. Chemosphere. 242:125145.

Goldner WS, Sandler DP, Yu F, Shostrom V, Hoppin JA, Kamel F, LeVan TD. (2013) Hypothyroidism and pesticide use among male private pesticide applicators in the agricultural health study. J. Occup. Environ. Med. 55(10):1171-8.

Hornick, J, Wasserman W (2012) Top ten reasons to shop at a farmers market. USDA. July 2, 2012.

Kondo M (2016) Sparking Joy. Ten Speed Press, Penguin Random House, LLC. New York.

Krueger F, Meyer-Lindenberg A. (2019) Toward a Model of Interpersonal Trust Drawn from Neuroscience, Psychology, and Economics. Trends Neurosci. 42(2):92-101.

Motta EVS, Raymann K, Moran NA. (2018) Glyphosate perturbs the gut microbiota of honey bees. Proc. Natl. Acad. Sci. USA. 115(41):10305-10310.

Roser-Renouf, C., Atkinson, L., Maibach, E., & Leiserowitz, A (2016). The Consumer as Climate Activist. Internatl. J. Commun. 10:24.

Saunders SP, Ries L, Oberhauser KS, Thogmartin WE, Zipkin EF (2018) Local and cross-seasonal associations of climate and land use with abundance of monarch butterflies Danaus plexippus. Ecography 41:278-290.

Sharma S, Chatterjee S. (2017) Microplastic pollution, a threat to marine ecosystem and human health: a short review. Environ Sci Pollut Res Int. 24(27):21530-21547.

Tosi S, Burgio G, Nieh JC (2017) A common neonicotinoid pesticide, thiamethoxam, impairs honey bee flight ability. Sci. Rep. 7:

Wang XT, Dvorak RD (2010) Sweet future: fluctuating blood glucose levels affect future discounting. Psychol. Sci. 21(2):183-8.

Yaumi AL, Abu Bakar MZ, Hameed BH (2017) Recent advances in functionalized composite solid materials for carbon dioxide capture. Energy 124:461-480.

Health-Promoting Nutrients and Phytochemicals

Micronutrient Information Center, Linus Pauling Institute (lpi. oregonstate.edu)

Fardet A, Boirie Y (2014) Associations between food and beverage groups and major diet-related chronic diseases: an exhaustive review of pooled/meta-analyses and systematic reviews. Nutrition Rev. 72(12):741-762.

ACKNOWLEDGMENTS

This book was previewed by a couple of eminent scientists.

Tom Lovejoy is the "Godfather of Biodiversity." He has served as an official scientific and environmental consultant for three US presidential administrations (Reagen, Bush, Clinton) and has worked tirelessly over five decades to preserve habitats and agriculture, primarily in South America. Few, if any, persons alive have done more to protect our planet from maladaptive human activities, and I am very lucky to work at the same university as Tom.

In my eyes, Robert Sapolsky is perhaps the best scientific storyteller on this planet. He served as my first postdoctoral mentor at Stanford University, taught me a new and healthier way to look at life, and continues to serve as a role model. These individuals generously provided intellectual support and/or critical feedback on the manuscript that helped to mold the overarching form of the book and the organization within the chapters.

Dr. Margaret Slavin, an associate professor in the Department of Nutrition and Food Studies at Mason, critically reviewed the health-promoting nutrients and phytochemicals list. She gently corrected my errors and provided an expert's insight into the current scientific understanding of how the foods we eat impact health and well-being.

Nhu Dumas supplied all of the traditional and non-traditional recipes at the end of each food chapter, she prepared all of the non-traditional dishes for photography, and she created the images of the prepared dishes. She also allowed me the time to read and research and to write this book by feeding, cleaning, teaching, and playing with our two little girls. Nhu is an amazing wife, mother, and chef.

CPSIA information can be obtained
at www.ICGtesting.com
Printed in the USA
LVHW071620210920
666680LV00026B/462